CW00523466

MICROBIOLOGY HANDBOOK

DAIRY PRODUCTS

Edited by

Rhea Fernandes

This edition first published 2008 by
Leatherhead Publishing, a division of
Leatherhead Food International Ltd
Randalls Road, Leatherhead, Surrey KT22 7RY, UK
URL: http://www.leatherheadfood.com

and

Royal Society of Chemistry
Thomas Graham House, Science Park, Milton Road,
Cambridge, CB4 0WF, UK
URL: http://www.rsc.org
Regstered Charity No. 207890

ISBN: 978-1-905224-62-3

A catalogue record of this book is available from the British Library

Printed by the MPG Books Group in the UK

FOREWORD

The Microbiology Handbook series includes Dairy Products, Fish and Seafood, and Meat Products, published by Leatherhead Food International and RSC Publishing. They are designed to provide easy-to-use references to the microorganisms found in foods. Each book provides a brief overview of the processing factors that determine the nature and extent of microbial growth and survival in the product, potential hazards associated with the consumption of a range of products, and growth characteristics for key pathogens associated with the product. All handbooks also contain a review of the related legislation in Europe and UK, guides to HACCP, and a detailed list of contacts for various food authorities. The books are intended to act as a source of information for microbiologists and food scientists working in the food industry and responsible for food safety, both in the UK and elsewhere.

Acknowledgements

The contributions of all members of staff at Leatherhead Food International who were involved with writing and reviewing the previous editions of this book are thankfully acknowledged. In the production of this edition, I would like to especially thank Dr Peter Wareing, Training Manager at Leatherhead Food International, for his valuable input into the book. His vast experience of food industry, and in specific 'food safety', has been priceless. I would also like to acknowledge Victoria Emerton, team leader for the technical team at Leatherhead Food International, for her careful editing; Eugenia Choi in our regulatory team who provided the update on legislation; Catherine Hill in our publications department for typesetting; and Ann Pernet for indexing. Finally, I am grateful to my parents, (late) Gabriel and Ana Fernandes, for all their encouragement and support over the years.

Rhea Fernandes
Leatherhead Food International

CONTENTS

INTRODUCTION

Milk and dairy products form a significant part of the human diet. They are rich sources of nutrients such as proteins, fats, vitamins and minerals; ironically, it is because of this that these products are susceptible to rapid microbial growth. In some instances, this microbial growth may be beneficial, while in others it is undesirable. Dairy products are vulnerable to spoilage or contamination with pathogens or microbial toxins; therefore, the microbiology of these products is of key interest to those in the dairy industry.

The Microbiology Handbook- Dairy Products consists of the microbiology of seven different dairy product categories: liquid milk products; concentrated and dried milk; cream, butter and spreads; cheese; fermented milks; and ice cream and frozen desserts, as well as HACCP. The third edition of this handbook provides a thorough review of the entire book for currency of information. Key changes in this edition are the recent regulatory changes pertaining to food hygiene and microbiological criteria for foodstuffs, and an emerging pathogen *Cronobacter sakazakii* (formerly known as *Enterobacter sakazakii*). This change in name was implemented in 2008, therefore all references published prior to 2008 will refer to the organism as *E. sakazakii*.

Further Reading

McSweeney P.L.H. The microbiology of cheese ripening, in *Cheese Problems Solved*. Ed. McSweeney P.L.H. Cambridge, Woodhead Publishing Ltd. 2007, 117-32.

Tamine A.Y., Robinson R.K. Microbiology of yoghurt and related starter cultures, in *Yoghurt: Science and Technology*. Eds. Tamine A.Y., Robinson R.K. Cambridge, Woodhead Publishing Ltd. 2007, 468-534.

Deak T. Yeasts in specific types of foods, in *Handbook of Food Spoilage Yeasts*. Ed. Deak T. Boca Raton, CRC Press. 2007, 117-201.

Hutkins R.W. Cultured dairy products, in *Microbiology and Technology of Fermented Foods*. Ed. Hutkins R.W. Oxford, Blackwell Publishing.2006, 107-44.

International Commission on Microbiological Specifications for Foods. Milk and dairy products, in *International Commission on Microbiological Specifications for Foods Microorganisms in Foods, Volume 6: Microbial Ecology of Food Commodities*. International Commission on Microbiological Specifications for Foods. London, Plenum Publishers. 2005, 643-715.

Walstra P., Wouters J.T.M., Geurts T.J. *Dairy Science and Technology.* Boca Raton, CRC Press. 2005.

Frohlich-Wyder M.-T. Yeasts in dairy products, in *Yeasts in Food: Beneficial and Detrimental Aspects.* Eds. Boekhout T., Robert V. Cambridge, Woodhead Publishing Ltd. 2003, 209-37.

Robinson R.K. *Dairy Microbiology Handbook.* New York, Wiley. 2002.

Marth E.H., Steele J.L. *Applied Dairy Microbiology.* New York, Marcel Dekker. 2002.

Fox P.F., Guinee T.P., Cogan T.M., McSweeney P.L.H. Pathogens and food-poisoning bacteria in cheese, in *Fundamentals of Cheese Science.* Eds. Fox P.F., Guinee T.P., Cogan T.M., McSweeney P.L.H. Gaithersburg, Aspen Publishers. 2000, 484-503.

Fox P.F., Guinee T.P., Cogan T.M., McSweeney P.L.H. Microbiology of cheese ripening, in *Fundamentals of Cheese Science.* Eds. Fox P.F., Guinee T.P., Cogan T.M., McSweeney P.L.H. Gaithersburg, Aspen Publishers. 2000, 206-35.

Fox P.F., Guinee T.P., Cogan T.M., McSweeney P.L.H. Bacteriology of cheese milk, in *Fundamentals of Cheese Science.* Eds. Fox P.F., Guinee T.P., Cogan T.M., McSweeney P.L.H. Gaithersburg, Aspen Publishers. 2000, 45-53.

Teuber M. Fermented milk products, in *The Microbiological Safety and Quality of Food, Volume 1.* Eds. Lund B.M., Baird-Parker T.C., Gould G.W. Gaithersburg, Aspen Publishers. 2000, 535-89.

Griffiths M.W. Milk and unfermented milk products, in *The Microbiological Safety and Quality of Food, Volume 1.* Eds. Lund B.M., Baird-Parker T.C., Gould G.W. Gaithersburg, Aspen Publishers. 2000, 507-34.

Neaves P., Williams A.P. Microbiological surveillance and control in cheese manufacture, in *Technology of Cheesemaking.* Ed. Law B.A. Sheffield, Sheffield Academic Press. 1999, 251-80.

Walstra P., Geurts T.J., Noomen A., Jellema A., van Boekel M.A.J.S. Microbiology of milk, in *Dairy Technology: Principles of Milk Properties and Processes.* Ed. Walstra P. New York, Marcel Dekker. 1999, 149-70.

Rampling A. The microbiology of milk and milk products, in *Topley and Wilson's Microbiology and Microbial Infections, Volume 2: Systematic Bacteriology.* Eds. Balows A., Duerden B.I. London, Arnold Publishers. 1998, 367-93.

International Dairy Federation, Jakobsen M., Narvhus J., Viljoen B.C. *Yeasts in the Dairy Industry: Positive and Negative Aspects; Proceedings of a Symposium,* Copenhagen, September 1996. IDF Special Issue No.9801. Brussels, International Dairy Federation. 1998.

Early R. *The Technology of Dairy Products.* London, Blackie. 1998.

Law B.A. *Microbiology and Biochemistry of Cheese and Fermented Milk.* London, Blackie. 1997.

International Dairy Federation. *The Significance of Pathogenic Microorganisms in Raw Milk.* Brussels, International Dairy Federation. 1994.

Varnam A.H., Sutherland J.P. *Milk and Milk Products: Technology, Chemistry and Microbiology.* London, Chapman and Hall. 1994.

Fox P.F. *Cheese: Chemistry, Physics and Microbiology, Volume 2: Major Cheese Groups.* London, Chapman and Hall. 1993.

Fox P.F. *Cheese: Chemistry, Physics and Microbiology, Volume 1: General Aspects.* London, Chapman and Hall. 1993.

Vasavada P.C, Cousin M.A. Dairy microbiology and safety, in *Dairy Science and Technology Handbook, Volume 2: Product Manufacturing.* Ed. Hui Y.H. Weinheim, VCH Publishers. 1993, 301-426.

White C.H., Bishop J.R., Morgan D.M. Microbiological methods for dairy products, in *Standard Methods for the Examination of Dairy Products.* Ed. Marshall R.T., American Public Health Association. Washington D.C, APHA. 1992, 287-308.

Flowers R.S., Andrews W., Donnelly C.W., Koenig E. Pathogens in milk and milk products, in *Standard Methods for the Examination of Dairy Products.* Ed. Marshall R.T., American Public Health Association. Washington D.C., APHA. 1992, 103-212.

Griffiths M.W., Stadhouders J., Driessen F.M. *Bacillus cereus* in liquid milk and other milk products, in *Bacillus cereus in Milk and Milk Products.* Ed. International Dairy Federation. Brussels, International Dairy Federation. 1992, 36-45.

McPhillips J., Smith G.J., Feagan J.T., Snow N., Richards R.J. The microbiology of milk: a review of growth of bacteria in milk and methods of assessment, in *Microbiology in Action.* Ed. Murrell W.G. Letchworth, Research Studies Press Ltd. 1988, 275-91.

Mabbitt L.A., Davies F.L., Law B.A., Marshal V.M. Microbiology of milk and milk products, in *Essays in Agricultural and Food Microbiology.* Ed. Norris J.R. Chichester, Wiley. 1987, 135-66.

1. LIQUID MILK PRODUCTS

1.1   Definitions

Milk is a complex biological fluid secreted in the mammary glands of mammals. Its function is to meet the nutritional needs of neonates of the species from which the milk is derived. This section of the handbook refers mainly to bovine milk, but the milk of other species, such as sheep and goats, is used for human consumption.

Typically, bovine milk is composed of approximately 87% water, 3.7 - 3.9% fat, 3.2 - 3.5% protein, 4.8 - 4.9% carbohydrate (principally lactose), and 0.7% ash. However, the exact composition of bovine milk varies with individual animals, with breed, and with the season, diet, and phase of lactation. Milk produced in the first few days post parturition is known as colostrum. Colostrum has a very high protein content, and is rich in immunoglobulin to help protect the newborn against infections. Colostrum is not generally allowed to enter the food supply in most countries.

Fresh milk products here refers to liquid milk, which accounts for about half of the total dairy market in the UK. Liquid milk is largely heat treated in developed countries, but a small quantity of raw (unpasteurised) milk is still sold in the UK. Skimmed and semi-skimmed milk, which are defined by their fat content (0.5%, and 1.5 - 1.8%, respectively), are increasingly important products in the liquid milk market.

1.2   Initial Microflora

1.2.1   Contamination from the udder

Although milk produced from the mammary glands of healthy animals is initially sterile, microorganisms are able to enter the udder through the teat duct opening. Gram-positive cocci, streptococci, staphylococci and micrococci; lactic acid bacteria (LAB), *Pseudomonas* spp. and yeast are most frequently found in milk drawn aseptically from the udder; corynebacteria are also common.

Where the mammary tissue becomes infected and inflamed; a condition known as mastitis, large numbers of microorganisms and somatic cells are usually shed into the milk. Mastitis is a very common disease in dairy cows, and may be present in a subclinical form, which can only be diagnosed by examining the milk for raised somatic cell counts. Many bacterial species are able to cause mastitis infection, but the most common are *Staphylococcus aureus, Streptococcus agalactiae, Streptococcus uberis* and *Escherichia coli*. These bacteria enter the udder by the teat duct, and *Staph. aureus* is able to colonise the duct itself. Although the organisms involved in mastitis are not usually able to grow in refrigerated milk, they are likely to survive, and their presence may be a cause of concern for health.

Diseased cows may also shed other human pathogens in their milk, including *Mycobacterium bovis, Brucella abortus, Coxiella burnetii, Listeria monocytogenes* and salmonellae. Recently, concerns have also been raised over the presence of *Mycobacterium avium* var. *paratuberculosis* (MAP) (the causative organism of Johne's disease in cattle) in milk from infected animals.

The outer surface of the udder is also a major source of microbial contamination in milk. The surface is likely to be contaminated with a variety of materials, including soil, bedding, faeces and residues of silage and other feeds. Many different microorganisms can be introduced by this means, notably salmonellae, *Campylobacter* spp., *L. monocytogenes*, psychrotrophic spore-formers, clostridia, and Enterobacteriaceae. Good animal husbandry and effective cleaning and disinfection of udders prior to milking are important in minimising contamination.

1.2.2   Other sources of contamination

Milking equipment and bulk storage tanks have been shown to make a significant contribution to the psychrotrophic microflora of raw milk if not adequately sanitised (1). Exposure to inadequately cleaned equipment and contaminated air are also sources of contamination (2). Milk residues on surfaces, and in joints and rubber seals can support the growth of psychrotrophic Gram-negative organisms such as *Pseudomonas, Flavobacterium, Enterobacter, Cronobacter, Klebsiella, Acinetobacter, Aeromonas, Achromobacter* and *Alcaligenes*, and Gram-positive organisms such as *Corynebacterium, Microbacterium, Micrococcus* and spore-forming *Bacillus* and *Clostridium* (3). These organisms are readily removed by effective cleaning and disinfection, but they may build up as biofilms in poorly cleaned equipment. Milk-stone, a mineral deposit, may also accumulate on inadequately cleaned surfaces, especially in hard water areas. Gram-positive cocci, some lactobacilli, and *Bacillus* spores can colonise this material and are then protected from cleaning and disinfection. Some of these organisms may survive pasteurisation and eventually cause spoilage (4).

Other, less significant, sources of contamination include farm water supplies, farm workers and airborne microorganisms.

1.2.3   Natural antimicrobial factors

Raw milk contains a number of compounds that have some antimicrobial activity. Their purpose is to protect the udder from infection and also to protect neonates, but they may also have a role in the preservation of raw milk during storage and transport.

Lactoperoxidase is an enzyme found in milk. It has no inherent antimicrobial activity, but, in the presence of hydrogen peroxide (usually of microbial origin), it oxidises thiocyanate to produce inhibitors such as hypothiocyanite. This is referred to as the lactoperoxidase system, and it has bactericidal activity against many Gram-negative spoilage organisms, and some bacteriostatic action against many pathogens. For this reason it has been investigated as a possible means of extending the life of stored milk (5)

Lactoferrin is also found in milk and is a glycoprotein that binds iron so that it is not available to bacteria. The chelation of iron in the milk inhibits the growth of many bacteria. In addition to producing an iron-deficient environment, lactoferrin is thought to cause the release of anionic polysaccharide from the outer membrane of Gram-negative bacteria, thereby destabilising the membrane.

Lysozyme acts on components of the bacterial cell wall, causing cell lysis. Gram-positive organisms are much more susceptible to lysozyme than Gram-negatives, although bacterial spores are generally resistant.

Immunoglobulins of maternal origin are often present in milk, and colostrum is a particularly rich source. These proteins may inactivate pathogens in milk, but their significance in preservation is uncertain.

1.3   Processing and its Effects on the Microflora

1.3.1   Raw milk transport and storage

In developed countries, raw milk on the farm is usually cooled quickly and stored in refrigerated bulk tanks at <7 °C prior to collection. Collection by insulated tanker is often on alternate days, or sometimes less frequently, and therefore some of the milk in the tank could be 48 hours old at the time of collection. Temperature control is thus critical to minimise microbial growth, and tanker drivers are usually permitted to refuse milk stored at too high a temperature, or which has an abnormal appearance or odour. Bacterial numbers in the milk may increase during transport, either as a result of contamination from inadequately cleaned tankers or from the growth of psychrotrophic organisms, particularly *Pseudomonas* spp.. Milk temperature and duration of the transport stage are therefore important factors.

On arrival at the processing site, the milk is transferred to bulk storage tanks, or silos, prior to processing. The milk may be stored in the silos for 2 - 3 days, and further growth of psychrotrophic bacteria is likely during this period. The degree of growth is dependent on the initial microbial load, and the storage time and temperature. Pseudomonads are the predominant organisms present in stored raw milk, with *Pseudomonas fluorescens*, *Pseudomonas fragi*, and *Pseudomonas lundensis* being commonly isolated (6), but Enterobacteriaceae, *Flavobacterium*, *Alcaligenes*, and Gram-positive species can also be found. The growth of psychrotrophic bacteria may also be accompanied by the production of heat-stable, extracellular proteolytic and lipolytic enzymes. These enzymes are often capable of surviving pasteurisation and, in some cases, ultra high temperature (UHT) processing, and they may subsequently cause spoilage in the processed milk.

A number of techniques have been used to limit the growth of psychrotrophs during raw milk storage.

1.3.1.1 Thermisation

The most commonly used technique is to apply a mild heat treatment (thermisation), by heating to around 57 - 68 °C for 15 - 20 seconds and then cooling rapidly to <6 °C. This reduces the psychrotrophic population significantly and can extend the storage life of the raw milk by several days. However, thermisation cannot eliminate vegetative pathogens, and is therefore not a reliable control for the hazard. For example, *L. monocytogenes* can survive the process and could then grow during chilled storage (7).

1.3.1.2 Deep cooling

As the storage temperature is a key factor for the rate of growth of psychrotrophic spoilage organisms, storing milk at as low a temperature as possible can also extend the storage life significantly. Reducing the storage temperature from 6 °C to 2 °C has been shown to give a 2-day gain in storage life for milk of good microbiological quality (8).

1.3.1.3 Carbon dioxide addition

There has been some interest in extending the storage life of raw milk by the addition of carbon dioxide at a concentration of 20-30 mM. Three mechanisms are thought to be involved in carbon dioxide inhibition of microorganisms: the first is by the displacement of oxygen; the second is a lowering of the pH of the milk due to the dissolution of carbon dioxide and formation of carbonic acid, particularly for Gram-negative psychrotrophic aerobes; and the third is a direct effect on the metabolisms such as inhibiting the production of enzymes by these organisms. It has also been suggested that the technique could be used to extend the shelf life of pasteurised milk, but concerns have been raised that the use of carbon dioxide addition could allow growth and toxin production by psychrotrophic *Clostridium botulinum*. However, recent work indicates that the risk of botulism is not increased by the use of this treatment (9).

Following storage, the milk then undergoes further processing.

1.3.2   Separation

If necessary, the milk is separated into skimmed milk, cream and sediment fractions, using centrifugal separators. The sediment may contain a comparatively high number of microorganisms and must be carefully discarded. The agitation involved may also break up clumps of bacteria, potentially producing an apparent increase in the number of colony-forming units. This process also allows the milk to be standardised to a specified fat content by adding back the correct quantity of cream.

1.3.3   Homogenisation

The fat globules in milk can coalesce and form a cream layer. Homogenisation reduces the size of the milk fat globules (to an average diameter of <1 μm) by using a pump to force milk through a valve under pressure. The fat globules are then small enough to remain in suspension. This process has little microbiological effect, although clumps of bacterial cells may be broken up. Homogenisers used for pasteurised milk may be linked to the pasteuriser, and run at raised temperature in order to minimise possible microbial contamination. UHT processed milks are homogenised in sterile conditions after heat treatment and before aseptic filling. Effective cleaning and sterilising of the homogeniser are then critical to product safety.

1.3.4   Pasteurisation

Some form of heat process is commonly applied to milk to ensure microbiological safety, and to extend shelf life. In the UK, the most commonly used process is pasteurisation. Time-temperature requirements for pasteurisation vary between countries, and are often specified in legislation. In the UK, both low-temperature, long time (LTLT, 63 - 65 °C for 30 minutes), and high-temperature, short time (HTST, 71.7 - 72 °C for at least 15 seconds) minimum processes are permitted. However, in practice, the HTST process is now generally used. Recent concern about the possible survival of MAP in pasteurised milk (discussed further in section 1.7.2.8: *MAP*) has seen many dairies increase the length of the HTST process to 25 seconds. Higher processes (such as ultra-pasteurisation at 138 °C for at least 2 seconds) (3) may also be applied to products with high fat and solids content. Plate heat exchangers are the most common method for milk pasteurisation, but it is essential that they are designed, constructed and operated in such a way as to minimise the possibility of recontamination of the pasteurised milk by raw milk. Most commercial pasteurisers are fitted with sensors that continuously monitor the pasteurisation temperature, and are linked to automatic divert valves. If the pasteurisation temperature falls below a specified value, the valve opens and diverts the under-processed milk away from the post pasteurisation section of the plant and the filling line, into a divert tank. The correct operation of these monitoring systems is critical and should be regularly checked. It is also essential that there are no cross-connections between the raw and pasteurised sides of the process, and this should include separate clean-in-place (CIP) systems. It is also usual to maintain a higher pressure in the pasteurised milk to minimise the risk of cross contamination in the heat exchanger. Recontamination of this kind may have serious public health consequences (discussed further in section *1.7.1: Pathogen growth and survival in raw milk*).

Accepted pasteurisation processes are designed to reduce the numbers of vegetative microbial pathogens to levels that are considered acceptable, although bacterial spores are not destroyed. Most of the potential psychrotrophic spoilage bacteria are also eliminated. However, certain heat-resistant mesophilic organisms, referred to as thermoduric, are able to survive pasteurisation. Thermoduric species commonly isolated from pasteurised milk include *Micrococcus* spp., *Enterococcus faecium* and *Enterococcus faecalis*, *Bacillus subtilis*, *Bacillus cereus*, and certain lactobacilli. Psychrotrophic strains of these organisms may be able to grow slowly in the pasteurised milk at 5 °C, and, if present initially in high numbers, could eventually cause spoilage. Effective cleaning of the cooling sections of pasteurisers is important to ensure that these organisms do not build up on surfaces.

1.3.5 UHT or sterilisation processes

Milk may also be subjected to more severe heat processes sufficient to achieve "commercial sterility". This may be done by batch heating in closed containers, or continuously with aseptic filling into sterile containers. Both conventional retort sterilisation and UHT processes must achieve a minimum F_0 of 3 minutes to ensure product safety. These processes destroy all vegetative cells in the milk, and the majority of spores, although certain very heat-resistant spores may survive. This results in a long shelf life without the need for refrigeration, but also causes organoleptic changes in the milk, such as browning.

Conventional sterilisation processes involve heating the milk in thick-walled glass bottles, closed with a crimped metal cap, at about 120 °C for approximately 30 minutes. However, modern large-scale production methods often use an initial UHT treatment prior to filling the container, followed by retorting for a reduced time (10 - 12 minutes), and then a rapid cooling process. This is said to give a product with improved organoleptic properties.

UHT processes may be direct or indirect. Direct systems inject high-pressure steam directly into the milk to obtain the desired temperature, and then employ flash cooling under vacuum to remove the resulting excess water. Indirect systems utilise heat exchangers and holding tubes. Direct systems are said to give better organoleptic properties, as the heating and cooling processes are very rapid, but they are more complex and expensive to install. UHT processed milk involves preserving milk by holding at a temperature of 140 - 150 °C for 1 - 2 seconds (minimum treatment is 130 °C for 1 sec) (3, 10). Heat treatment is usually followed by aseptic filling into sterile cartons or other containers. The maintenance of sterility in filling is vital to prevent recontamination of the treated milk. As with pasteurised milk, it is also vital to ensure that raw milk cannot recontaminate the UHT-treated milk.

Certain very heat-resistant spores of mesophilic bacilli, classified as *Bacillus sporothermodurans* (11) are able to survive UHT processes and may subsequently grow in the final product. However, this organism has been shown not to be pathogenic (12) and does not seem to cause any detectable changes to the product. Thermoduric *Bacillus stearothermophilus* are able to survive UHT processes and cause flat-sour spoilage (3).

1.4   Other Methods of Treating Milk

Because of the relatively short shelf life of conventional pasteurised milk, and the undesirable organoleptic changes in milk subjected to more severe heat processes, there has been much interest in alternative methods, both to improve product quality and to extend shelf life. Some of these processes are now being applied on a commercial scale in North America and Europe.

Microfiltration, usually using ceramic membrane filters, can be used in combination with a minimum HTST pasteurisation process to remove significant numbers of bacteria from milk, and give a substantial extension to shelf life over conventional pasteurised milk (13). The fat is separated from the milk before filtration and is heat treated separately before being added back to the milk after processing. Milk produced by this method is on sale in several countries, and is said to have a shelf life of at least 20 days.

Bactofugation is a centrifugation process that is also able to remove bacteria (including endospores) from milk. It has been used in the cheese industry for some years to minimise contamination with the spores of lactate-fermenting clostridia that cause 'late blowing'. The centrifugate produced by the process contains most of the microbial cells present initially in the milk, and this can be sterilised separately and then recombined with the treated milk, which is conventionally pasteurised, to restore its composition. A shelf life of 30 days or more is claimed for milk treated in this way.

Microwaving refers to dielectric heating due to polarisation effects at a selected frequency band (300 MHz to 300 GHz) in a nonconductor. It has been in commercial practice for milk pasteurisation for quite a long time as it provides the desired degree of safety with minimum quality degradation. Plate counts of raw milk undergoing continuous-flow microwave pasteurisation, at 2450 MHz, were negative while the temperature reached was 82.2 °C (14).

Other methods that have been applied to milk processing include irradiation, high-pressure processing, ultra sound treatment, ultra high-pressure homogenisation (UHPH), and pulsed-electric field (PEF).

1.5   Filling and Packaging

Although cleaning and hygiene of all processing equipment downstream of the heat treatment are essential to prevent recontamination of the product, for pasteurised milk it is the filling operation that is most likely to introduce microorganisms (15, 16). Psychrotrophic spoilage organisms may well be present on fillers, and these can then contaminate the milk and cause a significant reduction in shelf life. Microorganisms may also be present in the packaging, especially in poorly cleaned re-usable bottles, and this may also compromise the shelf life of the milk. For UHT-processed milk, aseptic filling into sterile containers is necessary for the maintenance of commercial sterility. Aseptic filling is not generally used for pasteurised milk, although it would be expected to have a significant influence on shelf life.

1.6 Spoilage

1.6.1   Pasteurised milk

Pasteurised milk provides a very suitable environment for microbial growth and is therefore highly susceptible to microbiological spoilage. Spoilage may result from either the growth of psychrotrophic thermoduric organisms that survive pasteurisation, or post-pasteurisation contamination by psychrotrophs. The latter is considered to be by far the most common cause of spoilage (17).

1.6.1.1   Thermoduric spoilage

The thermoduric microflora of milk consists largely of Gram-positive spore-formers, mainly *Bacillus* spp., *Clostridium* and organisms with heat-resistant vegetative cells, such as *Micrococcus, Lactobacillus, Enterococcus, Streptococcus, Corynebacterium* and *Alcaligenes*. Of these, the spore-formers are most important in spoilage, since the other species are not generally psychrotrophic and are unable to grow in refrigerated milk. Several *Bacillus* spp. contain psychrotrophic strains, notably *B. cereus* and *Bacillus circulans*, which may grow at temperatures as low as 2 °C. These organisms may become dominant in milk containing very low numbers of Gram-negative psychrotrophs, but even so, they rarely cause spoilage at <5 °C. However, at slightly higher temperatures (7 - 8 °C), *B. cereus* in particular may grow quite rapidly, producing a type of spoilage known as 'bitty cream' or 'sweet curdling', caused by the action of lecithinase on the phospholipids in fat globules. This produces small particles that stick to surfaces. Bitter taints may also be produced as a result of spoilage by *Bacillus* spp. These organisms are thought to originate from the raw milk, and the level of contamination has been shown to vary with the season, the highest numbers of spores being present between April and September (18).

1.6.1.2 Post-process contamination

The majority of post-process contaminants are Gram-negative bacteria, which may have some resistance to sanitisers and be able to colonise milk contact surfaces downstream of the pasteuriser. Initially, Enterobacteriaceae, such as *Enterobacter, Cronobacter,* and *Citrobacter*, predominate, but Gram-negative psychrotrophs, principally pseudomonads, but also *Alcaligenes, Klebsiella, Acinetobacter* and *Flavobacterium,* are more important in terms of eventual spoilage. Although these organisms may only contaminate the product in low numbers, they have a competitive advantage over Enterobacteriaceae at low temperatures and may grow rapidly to high levels (19). Spoilage by Gram-negative psychrotrophs usually takes the form of off-flavours, often described as unclean, fruity, rancid or putrid, formed as a result of proteolytic and lipolytic activity. Ropiness and partial coagulation may also occur occasionally. The time for spoilage to occur depends on the numbers and composition of the initial microflora, and the storage temperature (20). Milk produced with good hygienic practices in a modern facility should have a shelf life of more than 10 days at refrigerated storage temperatures.

Under conditions of mild temperature abuse, Enterobacteriaceae may predominate and cause acid clotting or the development of 'faecal' taints. At still higher temperatures, souring by LAB is possible.

Yeast and mould are also indicators of post-process contamination. Their presence and growth contribute to fruity and yeasty flavours in milk (2, 3).

1.6.2 *UHT or sterilised milk*

Spoilage of UHT-processed products is usually caused by post-process contamination. Spoilage caused by survival of heat-resistant *Bacillus* spores is rare, unless very large numbers of spores are present initially, although reports of sterility failure caused by *B. sporothermodurans*, as previously mentioned, are becoming more common.

Post-process contamination usually occurs as a result of a failure in the integrity of the aseptic filling system, or, more likely, as a result of packaging defects, such as pinholes or faulty seals. The product may then become contaminated with a variety of environmental organisms and the type of spoilage will be dependent on the nature of the contaminant. A spoilage rate of 1/10,000 units is a realistic target for producers using modern, well operated equipment.

A particular problem associated with UHT-processed milk is spoilage by heat-resistant, extracellular, proteolytic and lipolytic microbial enzymes. These will have been produced by psychrotrophic organisms growing in the raw milk prior to processing, particularly pseudomonads, *Acinetobacter*, and *Achromobacter*, which are then able to survive the thermal process, even though all viable cells have been destroyed. In the course of the long shelf life that these products are given, proteolytic enzymes can cause bitter flavours and gelation, whilst lipases cause the development of rancid flavours (21).

1.7 Pathogens: Growth and Survival

1.7.1 Raw milk

Before the adoption of routine pasteurisation, milk was an important vehicle for the transmission of a wide range of diseases, including typhoid, brucellosis and diphtheria. Pasteurisation and improvements in veterinary medicine have seen a very large reduction in the incidence of such traditionally milkborne diseases. However, raw milk may still contain a very wide range of pathogens, including *Salmonella* spp. (particularly *Salmonella typhimurium* and *Salmonella dublin*, a virulent serotype in humans), *E. coli* O157, *L. monocytogenes* and *Campylobacter* spp. derived from the milk animals, the environment or from farm workers and milking equipment (22). Pathogens may be present even in hygienically produced milk of generally good microbiological quality. In short, raw milk is a potentially hazardous product, the microbiological safety of which cannot be assured without the use of pasteurisation or an equivalent process. Recent milk-associated outbreaks of infectious intestinal disease in the UK have been shown to be caused mainly by unpasteurised or inadequately pasteurised milk products (23).

1.7.2 Pasteurised milk products

1.7.2.1 Salmonella

Salmonellae are not able to survive the typical minimum pasteurisation processes generally prescribed in legislation. Therefore, their presence indicates that the process has not been carried out effectively, or that post-process contamination has occurred. For example, an outbreak of salmonellosis in Kentucky in 1984 was associated with pasteurised milk, but an investigation of the dairy concerned showed that pasteurisation temperatures were inadequate, and could have been as low as 54.5 °C for 30 minutes (24). An outbreak caused by *Salmonella braenderup* in the UK in 1986 was also associated with pasteurised milk, and on this occasion the pasteuriser was found to be poorly designed and incorrectly operated, probably resulting in the application of an inadequate heat treatment (25).

In 1985, one of the largest outbreaks of salmonellosis in US history occurred in Illinois. Almost 200,000 people were affected, and were associated with pasteurised low-fat 2% milk contaminated with *S. typhimurium* (26). Investigations at the dairy plant involved revealed no evidence of inadequate pasteurisation, and the outbreak strain was not found to be abnormally heat resistant (27). Although the cause of the outbreak has never been completely explained, the investigation did discover a possible cross-connection between raw and pasteurised milk, which may have been the source of contamination (28).

In 1998, an outbreak of salmonellosis in Lancashire, caused by a multiresistant strain of *S. typhimurium* DT104, affected 86 people. This outbreak was also linked to defective pasteurisation of milk at a dairy on a local farm (29).

Consumption of raw milk or raw milk products have been responsible for 62 and 29 cases of diarrheal illness caused by *S. typhimurium* in 2003 and 2007, respectively, in the states of Ohio and Pennsylvania (30, 31)

Since salmonellae are occasional contaminants of raw milk, they may sometimes enter the processing environment. It is very important that contamination of the post-pasteurisation plant is not allowed to occur and effective precautions and monitoring procedures, based on HACCP principles, are necessary to prevent this.

1.7.2.2 Campylobacter spp.

Campylobacter spp. are not capable of surviving milk pasteurisation treatments, and cannot grow in raw or pasteurised milk, although they are able to survive for long periods in milk at refrigeration temperatures. Nonetheless, outbreaks of campylobacteriosis associated with pasteurised milk have occurred. For example, a large outbreak in the UK in 1979 caused by *Campylobacter jejuni* was estimated to have affected at least 2,500 schoolchildren, and was associated with free milk provided in schools. Although conclusive evidence was absent, it seems likely that raw milk may have bypassed the pasteurisation process (32).

A more recent outbreak in 2001 involved 75 people and was linked to the consumption of unpasteurised milk procured thorough a cow leasing program (33).

Birds are known to be an important reservoir of *Campylobacter* infection, and the tendency of some birds to peck through the foil tops of doorstep-delivered milk bottles is becoming recognised as an important source of infection in parts of the UK. Some individual cases have been attributed to this cause, and, in one instance in 1990, the organism was isolated from the beaks of jackdaws and magpies as well as the contaminated milk (34). More recently, an outbreak thought to be associated with bird-pecked milk was reported (35).

1.7.2.3 Listeria monocytogenes

There has been some discussion regarding the potential for *L. monocytogenes* in milk to survive pasteurisation. An outbreak of listeriosis in Massachusetts during 1983 resulted in 49 cases, 14 of whom subsequently died. Epidemiological evidence strongly suggested an association with consumption of pasteurised whole and low-fat (2%) milk, although this could not be confirmed microbiologically. The investigation failed to reveal any evidence of inadequate pasteurisation (a process of 77.2 °C for 18 seconds was applied), and the organism could not be found in environmental samples in the dairy, suggesting that post-process contamination was unlikely. However, samples of raw milk taken from farms supplying the plant were found to be positive for *L. monocytogenes* serotype 4b, and the investigators concluded that survival of some organisms through pasteurisation was the most likely cause of the outbreak (36). Three deaths and a miscarriage in Boston, USA between 2007-8 have been linked to presence of *Listeria* in pasteurised milk. So far, investigations have found nothing wrong with its pasteurisation process (37). Furthermore, in a survey of pasteurised milk conducted in Spain, *L. monocytogenes* was recovered from six out of 28 samples (21.4%) heated at 78 °C for 15 seconds (38). The explanation offered for both these findings was that the organisms might have been protected during heat treatment within leucocytes in the milk. However, this effect has not been conclusively demonstrated, and *L. monocytogenes* has not yet been shown to have survived pasteurisation in milk subjected to minimum HTST pasteurisation requirements of 71.7 °C for 15 seconds. For these reasons, it is currently accepted that existing pasteurisation processes are adequate to inactivate the organism in milk.

L. monocytogenes is likely to be present in wet dairy processing environments, and post-process contamination is therefore a particular hazard. The organism has been shown to be capable of significantly more rapid growth in pasteurised milk than in raw milk at 7 °C, and is also capable of growth at 4 °C in pasteurised milk (39). Therefore, effective HACCP-based controls to prevent post-process contamination are critical, particularly the cleaning and sanitising of all milk-contact surfaces. Adequate temperature control is also important.

1.7.2.4 *Verotoxigenic Escherichia coli*

Dairy cattle are an important reservoir for *E. coli* O157:H7 and this organism may therefore be present in raw milk, usually through faecal contamination. For this reason, raw milk is a high-risk food for this serious intestinal pathogen, and there have been a number of small outbreaks of infection associated with its consumption. However, *E. coli* O157:H7 is not a heat-resistant organism and there is no evidence that it is able to survive pasteurisation. Despite this, there have been outbreaks associated with pasteurised milk. In 1994, an outbreak in Scotland affected over 100 people and was associated with consumption of pasteurised milk from a local dairy. The outbreak strain was eventually recovered from cows on one of the farms supplying the dairy, from a bulk milk tanker, and from a pipe transferring milk from the pasteuriser to the bottling machine (40). Whether this outbreak was the result of faulty pasteurisation or post-process contamination was unclear, but, in either case, the raw milk is likely to have been the original source of the organism. In 1999, a serious outbreak occurred in Cumbria in the north-west of England, which was also associated with pasteurised milk from a local dairy. There were at least 60 confirmed cases involved, and the cause was thought to be a fault in the operation of the pasteuriser (41, 42).

The first general outbreak of verocytotoxin-producing *E. coli* (VTEC) in Denmark occurred in 2004 and involved 25 patients; 18 children and seven adults. It was thought to be due to the consumption of a particular kind of organic milk from a small dairy. Environmental and microbiological investigations at the suspected dairy did not confirm the presence of the outbreak strain, but the outbreak stopped once the dairy was closed and thoroughly cleaned (43).

E. coli O157 is not reported to be able to grow in raw or pasteurised milk stored at 5 °C, but may grow slowly at higher temperatures (44). However, since the infective dose of this pathogen is thought to be very low (probably fewer than 100 cells), effective pasteurisation and the prevention of post-process contamination are critical to ensure product safety.

1.7.2.5 *Yersinia enterocolitica*

Although there has been a question about the ability of *Y. enterocolitica* to survive milk pasteurisation, the majority of the evidence indicates that it is inactivated. Three different strains of *Y. enterocolitica* were reported to have D-values of 0.24 - 0.96 minutes at 62.8 °C (45). Therefore, the presence of the organism in pasteurised milk is likely to be the result of post-process contamination. There have been several *Y. enterocolitica* outbreaks associated with pasteurised milk. In 1976, an outbreak affecting 36 children was associated with the consumption of contaminated chocolate milk. It was thought that the organism was introduced to the product during mixing of chocolate syrup with pasteurised milk, without any subsequent heat process (46). Another outbreak in 1982 was the largest foodborne yersiniosis outbreak ever recorded in the USA, and was also associated with pasteurised milk. It is thought that several thousand people may have developed illness, although the organism was not isolated from milk or environmental samples at the dairy. It was found that surplus milk was used to feed pigs and that the crates used to transport this milk were stored on the ground at the farm and could have become contaminated with pig faeces. Since pigs are a well known reservoir for *Y. enterocolitica*, it was thought that inadequate washing of the crates allowed the organism to survive in mud on them, and subsequently contaminate the external surfaces of milk cartons (47).

Y. enterocolitica is capable of psychrotrophic growth, and could therefore multiply in pasteurised milk during storage. Measures should therefore be taken to prevent post-process contamination as with *L. monocytogenes*.

1.7.2.6 *Staphylococcus aureus*

Staph. aureus is only rarely involved in food poisoning associated with consumption of pasteurised milk, although enterotoxigenic strains can be found as contaminants in raw milk. This may be because *Staph. aureus* does not generally grow at temperatures below 7 °C, and enterotoxin production is inhibited at low temperatures. The organism is also known to be inhibited by the presence of competing species. Nevertheless, an outbreak in California affecting 500 school children was associated with chocolate-flavoured milk. The cause was thought to be growth of *Staph. aureus* in raw milk, and the subsequent persistence of the heat-stable enterotoxin through pasteurisation (48). In June and July 2000, a very large outbreak of staphylococcal food poisoning was reported in Japan, associated with consumption of pasteurised low fat milk. Over 14,500 people were said to have been affected. The outbreak was unusual in that the thermal processes had destroyed staphylococci in milk but *Staphylococcus* enterotoxin A (SEA) had retained enough activity to cause intoxication. SEA exposed at least twice to pasteurisation at 130 °C for 4 or 2 s retained both immunological and biological activities, although it had been partially inactivated (49).

1.7.2.7 Bacillus spp.

As has already been mentioned, psychrotrophic *Bacillus* spp. present in raw milk may survive pasteurisation and then become dominant in the pasteurised milk, potentially causing spoilage. Concerns have been expressed that some psychrotrophic strains of *B. cereus* may be able to produce toxin in milk at refrigeration temperatures, but it seems likely that obvious spoilage would occur before sufficient toxin production had taken place to cause illness (50). Even so, *B. cereus* was isolated at levels of 4x105 /g from pasteurised milk associated with 280 food poisoning cases in the Netherlands in 1989 (51).

1.7.2.8 Mycobacterium avium subsp. paratuberculosis

MAP is the causative organism of Johne's disease in cattle, a chronic wasting disease, and may occasionally be present in raw milk. Evidence linking MAP to a chronic inflammatory bowel condition in humans, called Crohn's disease, is becoming increasingly compelling. Concerns have been raised that MAP might be able to survive pasteurisation if present at levels above 100 cells per ml, especially if clumps of cells are present, and that pasteurised milk may therefore be a vehicle for Crohn's disease (52). On the basis of new heat-resistance studies, many UK dairies have increased pasteurisation times to from 15 to 25 seconds (53). A survey of the level of contamination of pasteurised milk by MAP over a 17 month period, in 1999 - 2000, revealed a mean of 1.6% of raw and 1.8% of pasteurised samples were positive for MAP cultures indicating that commercially pasteurised milk may occasionally contain low levels of viable MAP. The potential public health impact of this situation is, however, still uncertain given that an association with Crohn's disease in humans remains unproven (54).

1.7.2.9 Viruses

A number of viruses have been shown to be present in raw milk, although many of these, such as Foot and Mouth Disease Virus (FMDV), are not pathogenic to humans. However, raw milk has been implicated in outbreaks of hepatitis and poliomyelitis. Some viruses, including poliovirus, are completely inactivated by pasteurisation, but this seems not to be the case with others, such as FMDV, if the virus is naturally present rather than inoculated. There is therefore the possibility that other viruses pathogenic to humans may survive at low levels, but, in bulk milk processing systems, it is thought unlikely that sufficient viruses will be present to infect consumers (55).

1.7.2.10 Toxins

Mycotoxins may be present in milk as a result of the ingestion of mouldy and contaminated feed by cattle. Feed contaminated by aflatoxin B1 as a result of the growth of *Aspergillus flavus* or *Aspergillus parasiticus* has been shown to give rise to the presence of aflatoxin M1 in the milk of dairy cows consuming it. However, only a small percentage (0.4 - 2.2%) of the ingested toxin appeared in the milk (56). Aflatoxins are persistent compounds and are not greatly affected by milk processing, and could therefore be present in pasteurised, packaged milk. However, recent surveys suggest that contamination of the milk supply is very limited and well within acceptable levels (57).

1.8   References

1. McKinnon C.H., Rowlands G.J., Bramley A.J. The effect of udder preparation before milking and contamination from the milking plant on bacterial numbers in bulk milk of eight dairy herds. *Journal of Dairy Research*, 1990, 57, 307-18.

2. Eds. Doyle M., Beuchat L. *Food Microbiology: Fundamentals and Frontiers.* Washington, ASM Press. 2007.

3. Boor K., Fromm H. Managing microbial spoilage in the dairy industry, in *Food Spoilage Microorganisms*. Ed. Blackburn C. de W. Cambridge, Woodhead Publishing Ltd. 2006, 171-93.

4. Meer R.R., Baker J., Bodyfelt F.W., Griffiths M.W. *Psychrotrophic Bacillus* spp. in fluid milk products: a review. *Journal of Food Protection*, 1991, 54 (12), 969-79.

5. Wolfson L.M., Sumner S.S. Antibacterial activity of the lactoperoxidase system: a review. *Journal of Food Protection*, 1993, 56 (10), 887-92.

6. Ternstrom A., Lindberg A.-M., Molin G. Classification of the spoilage flora of raw and pasteurised bovine milk, with special reference to *Pseudomonas* and *Bacillus*. *Journal of Applied Bacteriology*, 1993, 75 (1), 25-34.

7. Mackey B.M., Bratchell N. The heat resistance of *Listeria monocytogenes*. *Letters in Applied Microbiology*, 1989, 6 (3), 89-94.

8. Griffiths M.W., Phillips J.D., Muir O.0. Effect of low-temperature storage on the bacteriological quality of raw milk. *Food Microbiology*, 1987, 4 (4), 285-91.

9. Hotchkiss J.H., Werner B.G., Lee E.Y.C. Addition of carbon dioxide to dairy products to improve quality: a comprehensive review. *Comprehensive Reviews in Food Science and Food Safety*, 2006, 5 (4), 158-68.

10. Eds. Jay J., Loessner M., Golden D. *Modern Food Microbiology*, New York, Springer. 2005.

11. Petterson B., Lembke F., Hammer P., Stackebrandt E., Priest F.G. *Bacillus sporothermodurans*, a new species producing highly heat-resistant endospores. *International Journal of Systematic Bacteriology*, 1996, 46, 759-64.

12. Hammer P., Suhren G., Heeschen W. Pathogenicity testing of unknown mesophilic heat resistant bacilli from UHT-milk, in *Bulletin of the International Dairy Federation*, No. 302. Ed. International Dairy Federation. Brussels, IDF. 1995, 56-7.

13. Eckner K.F., Zottola EA. Potential for the low-temperature pasteurisation of dairy fluids using membrane processing. *Journal of Food Protection*, 1991, 54 (10), 793-7.

14. Ahmed J., Ramaswamy H.S. Microwave pasteurisation and sterilisation of foods, in *Handbook of Food Preservation*, Ed. Rahman M.S. New York, Marcel Dekker. 2007, 691-711.

15. Moseley W.K. Pinpointing post-pasteurisation contamination. *Journal of Food Protection*, 1980, 43, 414.

16. Schroder M.J.A. Origins and levels of post pasteurisation contamination of milk in the dairy and their effects on keeping quality. *Journal of Dairy Research*, 1984, 51 (1), 59-67.

17. Champagne C.P., Laing R.R., Roy D., Mafu A. A, Griffiths M.W. Psychrotrophs in dairy products: their effects and their control. *CRC Critical Reviews in Food Science and Nutrition*, 1994, 34 (1), 1-30.

18. Phillips J.D., Griffiths M.W. Factors contributing to the seasonal variation of *Bacillus* spp. in pasteurised dairy products. *Journal of Applied Bacteriology*, 1986, 61 (4), 275-85.

19. Varnam A.H., Sutherland J.P. Liquid milk and liquid milk products, in *Milk and Milk Products: Technology, Chemistry and Microbiology*, Eds. Varnam A.H., Sutherland J.P. London, Chapman and Hall. 1994, 42-102.

20. Schroder M.J.A., Cousins C.M., McKinnon C.H. Effect of psychrotrophic post-pasteurisation contamination on keeping quality at 11 and 5 °C of HTST-pasteurised milk in the UK. *Journal of Dairy Research*, 1982, 49 (4), 619-30.

21. Law B.A. Reviews of the progress of dairy science: enzymes of psychrotrophic bacteria and their effects on milk and milk products. *Journal of Dairy Research*, 1979, 46 (3), 573-88.

22. Bryan F.L. Epidemiology of milk-borne disease. *Journal of Food Protection*, 1983, 46 (7), 637-49.

23. Djuretic T., Wall P.G., Nichols G. General outbreaks of infectious intestinal disease associated with milk and dairy products in England and Wales: 1992 to 1996. *CDR Review*, 1997, 7 (3), R41-5.

24. Adams D., Well S., Brown R.F., Gregorio S., Townsend L., Scags J.W., Hinds M.W. Salmonellosis from inadequately pasteurised milk: *Kentucky. Morbidity and Mortality Weekly Report*, 1984, 33, 504-5.

25. Rampling A., Taylor C.E.D., Warren R.E. Safety of pasteurised milk. *Lancet*, 1987, 2 (8569), 1209.

26. Ryan C.A., Nickels M.K., Hargrett-Bean N.T. Massive outbreak of antimicrobial-resistant salmonellosis traced to pasteurised milk. *Journal of the American Medical Association (JAMA)*, 1987, 258 (22), 3269-74.

27. Bradshaw J.G., Peeler J.T., Corwin J.J., Barnett J.E., Twedt R.M. Thermal resistance of disease-associated *Salmonella typhimurium* in milk. *Journal of Food Protection*, 1987, 50 (2), 95-6.

28. Lecos C. Of microbes and milk: Probing America's worst *Salmonella* outbreak. *Dairy and Food Sanitation*, 1986, 6 (4), 136-40.

29. Anon. Defective pasteurisation linked to outbreak of *Salmonella typhimurium* definitive phage type 104 infection in Lancashire. *CDR Weekly*, 1998, 8 (38), 335, 338.

30. Lind L., Reeser J., Stayman K., Deasy M., Moll W., Weltman A., Urdaneta V., Ostroff S., Chirdon W., Campagnolo E., Chen T. *Salmonella typhimurium* infection associated with raw milk and cheese consumption --- Pennsylvania, 2007, *MMWR Weekly*, 2007, (November 9), 56 (44), 1161-4.

31. Mazurek J., Salehi E., Propes D., Holt J., Bannerman T., Nicholson L.M., Bundesen M., Duffy R., Moolenaar R.L. A multistate outbreak of *Salmonella enterica serotype typhimurium* infection linked to raw milk consumption - Ohio, 2003. *Journal of Food Protection*, 2004, 67 (10), 2165-70.

32. Jones P.H., Willis A., Robinson D.A., Skirrow M.B., Josephs D.S. *Campylobacter enteritis* associated with the consumption of free school milk. *Journal of Hygiene*, 1981, 87 (2), 155-62.

33. Centres for Disease Control and Prevention. Outbreak of *Campylobacter jejuni* infection associated with drinking unpasteurised milk procured through a cow leasing program – Wisconsin, 2001. *Morbidity and Mortality Weekly Report*, 2002 / 51 (25), 548-9.

34. Hudson S.J., Lightfoot N.F., Coulson J.C, Russell K., Sisson P.R., Sobo A.O. Jackdaws and magpies as vectors of milkborne human *Campylobacter* infection. *Epidemiology and Infection*, 1991, 107 (2), 363-72.

35. Stuart J., Sufi F., McNulty C, Park P. Outbreak of *Campylobacter enteritis* in a boarding school associated with bird pecked bottle tops. *CDR Review*, 1997, 7 (3), R38-40.

36. Fleming D.W., Cochi S.L., MacDonald K.L., Brondum J., Hayes P.S., Plikaytis B.O., Holmes M.B., Audurier A., Broome C.V., Reingold A.L. Pasteurised milk as a vehicle of infection in an outbreak of listeriosis. *New England Journal of Medicine*, 1985, 312 (7), 404-7.

37. http://www.usatoday.com/news/nation/2008-01-08-Dairy-me_N.htm

38. Garayzabal J.F.F., Rodriguez L.D., Boland J.A.V., Cancelo J.L.B., Fernandez G.S. *Listeria monocytogenes* in pasteurised milk. *Canadian Journal of Microbiology*, 1986, 32 (2), 149-50.

39. Northolt M.D., Beckers H.J., Vecht V., Toepoel L., Soentoro P.S.S., Wisselink H.J. *Listeria monocytogenes*: heat resistance and behaviour during storage of milk and whey and making of Dutch types of cheese. *Netherlands Milk and Dairy Journal.* 1988, 42 (2), 207-19.

40. Upton P., Cola J.E. Outbreak of *Escherichia coli* O157 infection associated with pasteurised milk supply. *Lancet*, 1994, 344 (8928), 1015.

41. Anon. Outbreak of verocytotoxin producing *Escherichia coli* O157 infection in North Cumbria. *CDR Weekly*, 1999, 9 (11), 95-8.

42. Anon. VTEC 0157 phage type 21/28 infection in North Cumbria: update. *CDR Weekly*,1999, 9 (12), 108.

43. Eurosurveillance editorial team. Outbreak of verocytotoxin-producing *E. coli* O157:H7 linked to milk in Denmark. *Eurosurveillance Weekly*, 2004, 8 (20), 2465.

44. Wang G., Zhao T., Doyle M.P. Survival and growth of *Escherichia coli* O157:H7 in unpasteurised and pasteurised milk. *Journal of Food Protection*, 1997, 60 (6), 610-3.

45. Lovett J., Bradshaw J.G., Peeler J.T. Thermal inactivation of *Yersinia enterocolitica* in milk. *Applied and Environmental Microbiology*, 1982, 44 (2), 517-9.

46. Black R.E. Epidemic *Yersinia enterocolitica* infection due to contaminated chocolate milk. *New England Journal of Medicine*, 1978, 298 (2), 76-9.

47. Aulisio C.C, Lanier J.M., Chappel M. A. *Yersinia enterocolitica* 0:13 associated with outbreaks in three southern states. *Journal of Food Protection*, 1982, 45, 1263.

48. Evenson M.L., Ward Hinds M., Bernstein R.S., Bergdoll M.S. Estimation of human dose of staphylococcal enterotoxin A from a large outbreak of staphylococcal food poisoning involving chocolate milk. *International Journal of Food Microbiology*, 1988, 7 (4), 311-6.

49. Asao T., Kumeda Y., Kawai T., Shibata T., Oda H., Haruki K., Nakazawa H., Kozaki S. An extensive outbreak of *staphylococcal* food poisoning due to low-fat milk in Japan: estimation of enterotoxin A in the incriminated milk and powdered skim milk. *Epidemiology and Infection*, 2003, 130 (1), 33-40.

50. Langeveld L.P.M., van Spronsen W.A., van Beresteijn E.C.H., Notermans S.H.W. Consumption by healthy adults of pasteurised milk with a high concentration of *Bacillus cereus*: a double-blind study. *Journal of Food Protection*, 1996, 59 (7), 723-6.

51. van Netten P., van de Moosdijk A., van Hoensel P., Mossel D.A.A., Perales I. Psychrotrophic strains of *Bacillus cereus* producing enterotoxin. *Journal of Applied Bacteriology*, 1990, 69 (1), 73-9.

52. Hammer P., Knappstein K., Hahn G. Significance of *Mycobacterium paratuberculosis* in milk, in Bulletin of the International Dairy Federation, No.330. Ed. *International Dairy Federation*. Brussels, IDF. 1998, 12-6.

53. Grant I., Ball H., Rowe M. Effect of higher pasteurisation temperatures, and longer holding times at 72 °C, on the inactivation of *Mycobacterium paratuberculosis* in milk. *Letters in Applied Microbiology*, 1999, 28 (6), 461-5.

54. Grant I.R., Ball H.J., Rowe M.T. Incidence of *Mycobacterium paratuberculosis* in Bulk Raw and Commercially Pasteurised Cows' Milk from Approved Dairy Processing Establishments in the United Kingdom. *Applied and Environmental Microbiology*, 2002, 68 (5), 2428–35.

55. International Dairy Federation. Viruses pathogenic to man in milk and cheese, in *Behaviour of Pathogens in Milk. IDF Bulletin No. 122*. Ed. International Dairy Federation. Brussels, IDF. 1980, 17-20.

56. Frobish R.A., Bradley B.O., Wagner O.O. Long-Bradley P.E., Hairston H. Aflatoxin residues in milk of dairy cows after ingestion of naturally contaminated grain. *Journal of Food Protection*, 1986, 49 (10), 781-5.

57. Ryser E.T. Public health concerns, in *Applied Dairy Microbiology*. Eds. Marth E.H., Steele J.L. New York, Marcel Dekker. 2001, 397-546.

2. CONCENTRATED AND DRIED MILK PRODUCTS

2.1 Definitions

Concentrated and dried milk is produced for direct sale to the consumer, but are particularly important as food ingredients, providing a source of milk solids in a variety of other products. The removal of water from fresh milk gives advantages in terms of reduced storage and transport costs, convenience in use, and, in some cases, a useful extension to shelf life.

Concentrated milk is intended to be reconstituted by the consumer, by dilution with water, to give a similar composition to that of fresh milk. It is generally produced as a 3:1 concentrate containing approximately 10 - 12% milk fat and 36% total milk solids. It gives little benefit in terms of keeping quality or convenience, and has not become an important product.

Bulk condensed milk is an important source of milk solids in confectionery, bakery products, ice cream, concentrated yoghurt and other products, and is manufactured in large quantities for this purpose. It may be made from whole, skimmed, or reduced fat milks, depending on the end use. Most bulk condensed milk is made by evaporation, and the degree of concentration is usually within the range 2.5:1 to 4:1, depending on usage. The keeping quality is limited because it is not sterilised during or after processing.

Sweetened condensed milk may be made from whole or skimmed milk either in bulk as a food ingredient, or in small cans or tubes for direct sale to the consumer. When made from whole milk, it should contain at least 8% fat and 28% total milk solids; when made from skim milk it contains at least 0.5% fat and 24% milk solids. After evaporation, sufficient sugar is added, usually as sucrose or glucose, to prevent most microbial growth. The sugar concentration in the aqueous phase (known as sugar ratio, sugar number or sugar index) of retail products is usually in the range of 63.5 - 64.5, giving a water activity (a_w) of about 0.86, but is often much lower in bulk material. For bulk, whole milk products the sugar index is about 42; this is because the product is stored refrigerated for fairly short periods. The product is packed in hermetically sealed metal containers for retail trade, and milk cans, barrels, steel drums or bulk tanks for industrial purposes. Significant quantities of bulk sweetened condensed milk are used in the confectionery, bakery and prepared food industries.

Evaporated milk is similar in composition to bulk condensed whole milk, but is usually heat processed and canned to give a 'commercially sterile' product.

Evaporated milk is produced by the removal of about 60% water from whole milk, which results in the lactose content being about 11.5%. It normally contains at least 7.5% fat and 25 - 26% total milk solids, and is also permitted to contain stabilising salts to maintain optimum viscosity after sterilisation. Some products are made from skimmed milk, and some are 'filled milk' where other fats are used to replace milk fats. Most evaporated milk is sold directly to consumers for use in home cooking.

Dried milks are manufactured as food ingredients in large quantities, and with widely varying compositions, depending on their end use. Both dried whole milk and skimmed milk powders are produced, but there is also a range of products with specific characteristics for particular applications in food processing e.g. lipase-free, calcium-reduced. Dried milks are now mostly produced by spray-drying, or occasionally by drum or roller-drying, and usually have a moisture content of less than 5% to give microbiological stability.

2.2   Initial Microflora

The initial microflora of concentrated milk is that of the raw milk from which it is produced. Sugar used in sweetened condensed milk may be an additional source of yeasts and moulds and bacterial spores, including thermophilic spores.

2.3   Processing and its Effects on the Microflora

Examples of processes used to produce condensed, evaporated and dried milk products are shown in Figures 2.1 - 2.3. In each case, the initial steps of milk storage, transport, separation and standardisation are the same as those used for fresh milk. The first key stage is therefore the pasteurisation or pre-heating process. The time and temperature applied depends on the intended use of the product, but, as a minimum, will correspond to milk pasteurisation (72 °C for 15 seconds). Much higher processes are used in some instances. For example, in large-scale continuous production of evaporated milks, temperatures as high as 121 °C may be applied for several minutes to stabilise milk proteins. Condensed milk is also subjected to processes more severe than milk pasteurisation, with the exact process being determined by the nature of the product required. This pre-heating helps to increase viscosity and improve other characteristics. These processes may be less controlled than conventional pasteurisation, but it is important that a safe minimum process is always applied. As with fresh milk, most vegetative bacteria will be destroyed during heating, but some thermoduric types and bacterial endospores are likely to survive all but the most severe processes.

The second common process in the manufacture of all types of concentrated milk products is the removal of water, usually by evaporation. The most common type of evaporator used in the dairy industry is the falling film evaporator, which is both energy-efficient and readily controllable. It is common to link several evaporators together in series to form what is known as a 'multiple effect evaporator', with a common condenser and vacuum source. The vapour produced in the first effect heats the second, and so on, producing a stepwise decrease in temperature from 70 - 80 °C in the first effect, to about 40 °C in the last. The vacuum, however, is greater in the lowest temperature effect, so that the milk flows from high to low temperature. This process is very efficient and produces milk at the required concentration without a second pass. The temperatures within the lower temperature effects of the evaporator are low enough to permit the growth of thermophilic and some mesophilic spores, and quite high numbers may develop in the effects during prolonged production runs. Growth of certain thermophilic species has even been reported in milk at temperatures as high as 70 °C (1). The growth of thermophiles must be controlled, by limiting the length of production runs, effective plant cleaning and sanitation, and ensuring that adequate standards of plant hygiene are applied.

2.3.1 Concentrated milk

Raw milk is given a heat treatment approximating that of pasteurisation. It is then concentrated at a low temperature, followed by standarisation, homogenisation and pasteurisation before packaging. Pasteurisation is usually done at 79.4 °C for 25 sec; the product is stored at 10 °C to prevent growth of thermoduric bacteria and any post-pasteurisation contaminants (2).

2.3.2 Bulk condensed milk

In the manufacture of bulk condensed milk, the milk is first separated, if necessary, and standardisation of fat content is often carried out after concentration. Unless skimmed milk is used, homogenisation is also usually carried out at this stage. A pre-heating process is then applied, using a continuous heater or a 'hot well'. Temperatures of 65.6 – 76.7 °C are often used, but higher temperatures of 82.2 - 93.3 °C for as much as 15 minutes may be applied to obtain a higher viscosity product and to impart other desirable characteristics. These processes will greatly reduce the number of vegetative bacterial cells in the milk, but they are not under the same degree of control as conventional pasteurisation. The heater or 'hot wells' can act as incubators for thermophilic bacteria, especially if the product is held in the lower temperature range, allowing bacterial numbers to build up; it is also possible that pathogens may survive some lower pre-heat temperatures. Therefore, a further HTST pasteurisation step should be applied to eliminate this hazard. The preheated milk is then concentrated in a vacuum-pan or in a multiple-effect evaporator at a temperature range of 54.4 - 57.2 °C After evaporation, the product is not sterile and will support rapid microbial growth. It is also likely that post-process contamination could occur during standardisation or packaging. Therefore, effective post-process hygiene procedures, and rapid cooling to <5 °C are necessary to achieve the required shelf life. The process is depicted in Figure 2.1.

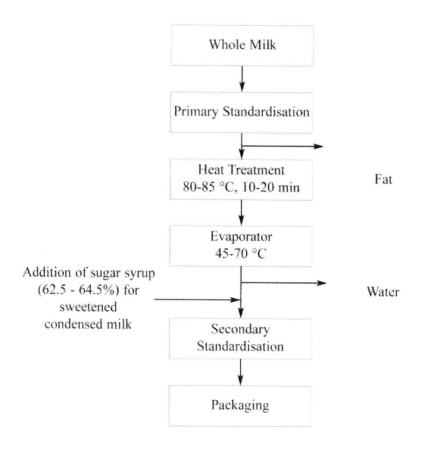

Fig. 2.1. Production of condensed milk

2.3.3 *Sweetened condensed milk*

In the manufacture of sweetened condensed milk, the milk is pre-heated, and may also undergo superheating. Temperatures within the range 82 - 100 °C for 10 - 30 minutes are used, and these processes are sufficient to destroy vegetative cells. For bulk, sweetened condensed milk, sugar can be added before concentration at varying levels depending on the end use. This material is not usually microbiologically stable and has a limited shelf life. In the production of sweetened condensed milk for retail sale, sugar (generally sucrose) is added during the later stages of evaporation. The sugar is normally added as a 65% solution, and the finished product then contains a high enough solute concentration to give an a_w of 0.83 - 0.86. It is this reduced a_w, rather than a heat process, that confers microbiological stability, and a long shelf life when packed in pre-sterilised cans or tubes. However, some osmophilic yeasts and moulds are able to grow in the product, and may cause spoilage. Evaporation is carried out at a temperature of around 57.2 °C, but can drop to 48 - 9 °C late in the cycle. Partially cooled milk (30 °C) is seeded with very fine lactose crystals to force crystallisation.

A high standard of hygiene during the later handling and filling stages is necessary to prevent contamination. A particular problem with sweetened condensed milk is its high viscosity, and 'sticky' nature, which makes cleaning processing equipment difficult. This high viscosity means that positive-displacement, plunger-type fillers are needed to fill the product. These are complex and difficult to clean and may become heavily contaminated with micrococci and yeasts if not properly maintained.

2.3.4   Evaporated milk

The manufacture of canned evaporated milk is very similar to that of bulk condensed milk, but the product is given a long shelf life by applying a heat process designed to give commercial sterility. Because of this, the milk has to be stabilised to prevent coagulation during processing and to minimise 'age thickening' during storage. Stabilisation is achieved by the addition of permitted salts, including phosphates, citrates and bicarbonates, which are used to maintain the pH of the milk at 6.6 - 6.7. Pre-heating is also important, and temperatures of 120 - 122 °C for several minutes are used to denature whey proteins. Only a few bacterial spores are likely to survive such a process. Condensation is usually performed at a temperature lower than 54.5 °C (2).

After evaporation, the milk is homogenised, cooled and stored. It is then standardised, and further stabilising salts may also be added at this point. It is important to ensure that cooling is rapid and sufficient to minimise any microbial growth during this procedure. The milk is then filled into cans, hermetically sealed, and sterilised using batch retorts or continuous sterilisers. Processing at 115 °C for 15 - 20 minutes, or 120 °C for 10 minutes, has been traditionally applied, but recently there has been a move towards UHT processing followed by aseptic filling into pre-sterilised cans or cartons. Processing at 130 °C for 30 seconds, to 150 °C for less than one second may be used. Retorted canned milk is commercially sterile, and only extremely heat-resistant spores of organisms such as *Bacillus stearothermophilus* are likely to survive. These spores do not germinate unless the cans are then stored at high ambient temperatures (>43 °C). Aseptically filled evaporated milk may become recontaminated during filling, unless stringent hygiene procedures equivalent to those used in filling other UHT milk products are used. The process is depicted in Figure 2.2.

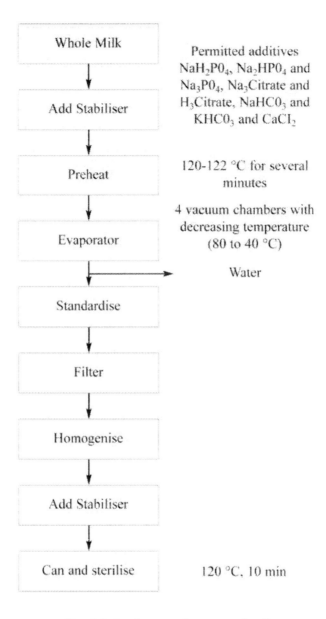

Fig. 2.2. Production of evaporated milk

2.3.5 *Dried milk*

Drying is used, after evaporation, to produce a stable, powdered product with a residual moisture content of 2 - 5%. Most dried milk powders are now spray-dried, and drum and roller dryers are little used. Spray drying is a more energy-efficient process and causes less heat damage to the product. Therefore, only spray drying will be considered further.

Before evaporation and drying, the milk is standardised, if required, and heat-treated. Vegetative bacteria, including Enterobacteriaceae and *Listeria monocytogenes* (3, 4) have been shown to survive the drying process, and therefore all raw milk should receive a process at least equivalent to pasteurisation. Skimmed milk may also be subjected to low, medium and high heat processes to give varying degrees of protein denaturation as required. Typically, a low heat process will be 74 °C for 30 seconds, a medium heat process 80 - 100 °C for 1 - 2 minutes, and a high heat process may be equivalent to an ultra high temperature (UHT) treatment (140 - 150 °C). Milk for dried whole milk powders is heated at 85 - 95 °C for several minutes. UHT heating units result in products that have excellent microbiological quality, which is important when dried milk is to be used as an ingredient in baby food. The heat-treated milk is then evaporated to about 50% total solids before drying. As the drying process does not kill all vegetative bacterial cells, effective cleaning and hygiene procedures are required to ensure that the heat-treated milk does not become recontaminated.

The current practice is to link the evaporation plant and dryer as a single integrated unit. Two systems of spray drying are used: jet or nozzle dryers, and rotary atomiser dryers. Each dryer produces powders with specific characteristics. In the case of jet dryer systems, milk is fed to a high-pressure pump, then pumped under pressure to a series of jets or nozzles within the drying chamber. The impact of hot air on the milk causes the liquid to break up into fine droplets that form milk powder. In the rotary atomiser, the milk enters the drying chamber through slots or holes located at the periphery of a circular disc fixed to a rotating shaft.

Modern spray dryers usually consist of several stages, and incorporate a final drying stage using a fluidised bed dryer.

Spray dryers operate by mixing pre-heated atomised milk droplets with heated air at an inlet temperature of 180 - 220 °C. The air is cooled to an exit temperature of 70 - 95 °C, and moisture is removed from the milk. This gives very rapid drying at a vaporisation temperature of about 50 °C, and the temperature of the dried particles remains below that of the cooled air. The dried particles are then removed, cooled, and packaged. The a_w of the finished milk powder (0.3 - 0.4) is too low to allow any microbial growth. The full process is depicted in Figure 2.3.

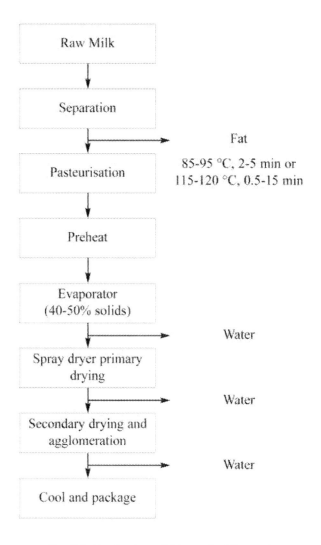

Fig. 2.3. Production of skimmed milk powder

2.4   Spoilage

2.4.1   *Condensed milk*

Unsweetened condensed milks are not commercially sterile, and so, are favourable media for microbial growth. Spoilage can be caused by heat-resistant organisms from raw milk, for example *Bacillus* spp. and enterococci, or by post-process contaminants, such as pseudomonads and members of the Enterobacteriaceae. As long as the product is handled and stored correctly (<7 °C) only thermoduric and thermophilic organisms will grow slowly (5). Shelf life varies from a few days to weeks, depending on the degree of contamination, the severity of the heat treatment applied, and the effectiveness of temperature control during cooling and storage. The pattern of spoilage is very similar to that described for pasteurised fresh milk, although organisms adapted to slightly lower aw values may have an advantage.

2.4.2 Sweetened condensed milk

The low a_w (0.85) of sweetened condensed milks ensures that only osmophilic and osmotolerant organisms are able to grow. Canned products may be spoiled by slow growth of osmophilic yeasts, particularly *Torulopsis* spp., which enter the product after heating and may produce sufficient gas to cause blown cans (2). If sufficient oxygen is present in the headspace, or the can has a small pinhole leak, moulds such as *Aspergillus* and *Penicillium* spp. may grow as 'buttons' on the surface of the product. This problem is associated with poor plant hygiene (2, 6).

Bulk products with lower sugar concentrations are more susceptible to spoilage. Mould growth may occur on the surface of stored milk. Again, *Penicillium* and *Aspergillus* are the main genera involved. Occasionally, bacterial spoilage by osmotolerant micrococci and *Bacillus* spp. may occur in these products, causing thickening, and eventually, lipolysis and proteolysis (7).

2.4.3 Evaporated milk

Spoilage of commercially sterile canned evaporated milk is uncommon, but is a result of either under-processing, or post-process contamination. Species of bacilli such as *B. stearothermophilus*, *Bacillus coagulans*, and *Bacillus licheniformis* may survive the heat process and cause acid coagulation, a slight cheesy odour and flavour, and 'flat sour' spoilage of the milk (8). However, many strains are obligately thermophilic and are only a problem at elevated storage temperatures, or if cooling is too slow. Non-acid curd is caused by *Bacillus subtilis*; this can then be digested to a brownish liquid with a bitter taste. *Bacillus megaterium* is responsible for the formation of coagulum, which is accompanied by cheesy odour and gas (2). Blown cans associated with putrefactive spoilage by *Clostridium* spp. occur very occasionally. Spoilage caused by under-processing is very rare in recent times, mainly as a result of improvements in process technology and control.

Post-process spoilage can be caused by a wide range of species, and the spoilage characteristics are therefore varied. The situation is very similar to that described for other sterilised, and UHT milk products. Bacteria may enter the product as a result of faulty can seaming, or subsequent seam damage, corrosion, and in aseptically filled products, following a breakdown in the integrity of the aseptic filling process.

2.4.4   Dried milks

Relatively low numbers of microorganisms survive processing. Heat resistant organisms (spore-formers and non-spore-formers) and mould are responsible for deterioration of milk powders, if the product is allowed to absorb moisture during prolonged storage. The a_w of dried milks is otherwise much too low to support microbial growth, and a general decrease in microbial counts occurs during storage (5).

2.5   Pathogens: Growth and Survival

Concentrated milks are not normally regarded as high-risk products, principally because of the relatively severe heat treatments used in their manufacture. As with other pasteurised and UHT processed milks, the main concern for condensed and evaporated milk is post heat-treatment contamination by pathogens.

Dried milks, however, are the subject of considerable concern, particularly in view of their widespread use in infant foods. There have been a number of outbreaks of foodborne disease associated with dried milk powders.

2.5.1   Condensed/evaporated milk

2.5.1.1   Listeria spp.

The fate of *L. monocytogenes* has been studied in these products. The organism declined during storage in sweetened condensed milk at 21 °C, but the population remained stable at 7 °C. In evaporated milk, growth was recorded at both temperatures (9).

2.5.1.2   Clostridium spp.

A study of the incidence of clostridia in sweetened condensed milk showed that about 40% of the samples contained >100 cfu/100 g. These contaminants were identified mainly as *Clostidium butyricum* and *Clostridium perfringens* (10). However, the a_w of these products is too low to allow the germination of spores and vegetative cell growth.

2.5.1.3   Staphylococcus aureus

Although there are no reported cases of foodborne disease associated with canned sweetened condensed milk, its a_w of 0.85 is very close to the minimum value that would allow *Staph. aureus* to grow, although toxin production would be inhibited. However, bulk products with much lower sugar contents might be at risk if they become contaminated. Therefore, adequate hygiene is an important control.

2.5.2   Dried milk

Although dried milk products have been implicated in a number of foodborne disease outbreaks, these have usually been the result of post-pasteurisation contamination by pathogens. Foodborne pathogenic bacteria are unable to grow in dried milk powders, but may survive for long periods.

2.5.2.1   Salmonella spp.

There have been several significant salmonellosis outbreaks associated with dried milk powders, and *Salmonella* contamination has come to be regarded as a serious potential hazard in these products. In 1964 to 1965, a nationwide outbreak occurred in the USA associated with non-fat milk powder produced at one plant, but then agglomerated (instantised) at a number of other locations (11). This outbreak produced reports of infection throughout the USA and led to a major United States Department of Agriculture (USDA) investigation of the incidence of *Salmonella* contamination in milk drying plants. It was found that contamination was widespread in both product and environmental samples, and this finding gave rise to a number of improvements in hygiene, sanitation and process control. Despite this, another smaller outbreak occurred in Oregon in 1979, associated with non-fat dried milk contaminated with *Salmonella typhimurium* and *Salmonella agona*. Surveillance results have also continued to show persistent low-level (<1% of samples) *Salmonella* contamination in the US.

In 1986, a major outbreak of salmonellosis was reported in the UK associated with dried milk powder-based infant formula, contaminated with *Salmonella ealing*. The organism is thought to have originated in raw milk and then spread through the plant. It seems to have entered the insulation material of the dryer through small cracks in the dryer wall, and this then acted as a reservoir from which *S. ealing* could repeatedly contaminate the finished product (12). In 2005, powdered infant formula contaminated with *S. agona* was implicated in an outbreak involving 104 infants in France (13). Despite the introduction of HACCP and further recommendations for the prevention of contamination (14), dried milk-associated outbreaks continue to be reported around the world (15, 16). Such outbreaks are often linked to contamination in equipment that is poorly designed and difficult to clean effectively, and the *Salmonella* strains involved are often found to be lactose-positive.

2.5.2.2   Staphylococcus aureus

Contamination of dried milk powders with staphylococcal enterotoxins was a significant problem in the 1950s, and several outbreaks were recorded, often caused by growth and toxin production in the concentrated milk prior to drying (17, 18). Improvements in temperature control and hygiene prior to drying have largely eliminated this problem.

However, in 1986, several outbreaks were reported in Egypt associated with imported non-fat dried milk. Analysis of samples showed no viable pathogens, but staphylococcal enterotoxins A and B were found at concentrations high enough to cause illness (19). More recently, in 2000, a very large outbreak of staphylococcal food poisoning was reported in Japan, which affected over 13,420 people. This outbreak was associated with consumption of semi-skimmed liquid milk, which was manufactured using dried skimmed milk powder. It was thought that some temperature abuse could have occurred during production of the dried milk, allowing *Staph. aureus* to grow and produce heat-stable toxin, which then persisted through to the finished product and cause intoxication, even though the thermal processes had destroyed the organism (20).

2.5.2.3 Listeria monocytogenes

No cases of listeriosis associated with dried milk products have been reported. However, the ubiquity of *Listeria* spp. in dairy plants and other wet processing areas, and the cases of listeriosis linked to other dairy products suggest that contamination of dried products is likely. The survival of *L. monocytogenes* during spray drying and storage of product has been investigated. Spray drying was found to give a small reduction in numbers, and the viable count continued to decline during storage, but viable *L. monocytogenes* could still be isolated from some samples after 12 weeks (4).

2.5.2.4 Bacillus spp.

Bacillus cereus has been found to be a common contaminant in dried milk. In the US, 62.5% of samples of milk powder were found to be positive, and, in Brazil, the organism was isolated from 80% of samples examined (21). Although there have been many reports of *B. cereus* food poisoning associated directly with dried milk consumption, in 2005, milk powder contaminated with *B. licheniformis* and *B. subtilis* was the cause of an outbreak in Croatia involving 12 children. Reconstituted milk that was held for 2 hours prior to consumption, without boiling, was identified as the cause (22). *B. cereus* spores can survive for many months in dried milk powders, and rapid growth has been shown in reconstituted powders at ambient temperature (21). Dried milk is thought to have been the source of *B. cereus* in an outbreak affecting eight people, associated with macaroni cheese, which was found to contain *B. cereus* at levels of 10^8 - 10^9 cfu/g (23).

2.5.2.5 Cronobacter and Enterobacter spp.

Cronobacter and *Enterobacter* spp. are not normally regarded as foodborne pathogens, but there have been a number of sporadic outbreaks of neonatal meningitis caused by *Cronobacter* (*Enterobacter*) *sakazakii* associated with dried milk consumption, with fatality rates as high as 30 – 80% (24, 25, 26). Powdered infant formulae contaminated with *C. sakazakii* was responsible for outbreaks among infants: one involving nine infections and two deaths in 2004, in France, and the other caused five infections and one death in 2005, in New Zealand (13). These outbreaks are thought to have been due to growth of the organism in the reconstituted powder. The presence of any members of the Enterobacteriaceae in infant formulae may therefore be a cause for concern.

Studies have shown that *C. sakazakii* can survive spray-dying when inoculated into skimmed milk powder (27). *Enterobacter agglomerans* has also been isolated from milk powder (28).

2.5.2.6   Toxins

The mycotoxin aflatoxin Ml has occasionally been found in dried milk (29). The drying process has been found to reduce the concentration, but a significant amount is able to survive processing and storage of finished product for long periods (30).

2.6   References

1. Langeveld L.P.M., van Montfort-Quasig R.M.G.E., Weerkamp A.H., Waalewijn R., Wever J.S. Adherence, growth and release of bacteria in a tube heat exchanger for milk. *Netherlands Milk and Dairy Journal*, 1995, 49 (4), 207-20.

2. Robinson R., Itsaranuwat P. The Microbiology of Concentrated and Dried Milks, in *Dairy Microbiology Handbook: The Microbiology of Milk and Milk Products*. Ed. Robinson R., New York, John Wiley & Sons, Inc. 2002, 175-212.

3. Daemen A.L.M., van der Stege H.J. The destruction of enzymes and bacteria during the spray drying of milk and whey. 2. The effect of the drying conditions. *Netherlands Milk and Dairy Journal*, 1982, 36, 211-29.

4. Doyle M.P., Meske L.M., Marth E.H. Survival of *Listeria monocytogenes* during the manufacture and storage of nonfat dry milk. *Journal of Food Protection*, 1985, 48 (9), 740-2.

5. Clarke W. Concentrated and Dry Milk and Wheys, in *Applied Dairy Microbiology*. Eds. Marth E., Steele J.. New York, Marcel Dekker, Inc. 2001, 77-92.

6. International Commission on Microbiological Specifications for Foods. Milk and dairy products, in *International Commission on Microbiological Specifications for Foods Microorganisms in Foods, Volume 6: Microbial Ecology of Food Commodities*. Ed. International Commission on Microbiological Specifications for Foods. London, Plenum Publishers. 2005, 643-715.

7. Varnam A.H., Sutherland J.P. Concentrated and dried milk products, in *Milk and Milk Products: Technology, Chemistry and Microbiology*. Eds. Varnam A.H., Sutherland J.P. London, Chapman and Hall. 1994, 103-58.

8. Kalogridou-Vassiliadou D. Biochemical activities of *Bacillus* species isolated from flat sour evaporated milk. *Journal of Dairy Science*, 1992, 75 (10), 2681-6.

9. Farrag S.A., EI-Gazzar F., Marth E.H. Fate of *Listeria monocytogenes* in sweetened condensed and evaporated milk during storage at 7 or 21 °C. *Journal of Food Protection*. 1990, 53 (9), 747-50, 770.

10. Bhale P., Sharma S., Sinha R.N. Clostridia in sweetened condensed milk and their associated deteriorative changes. *Journal of Food Science and Technology*, 1989, 26 (1), 46-8.

11. Collins R.N., Trager M.D., Goldsby J.B., Boring J.R., Cohoon D.B., Barr R.N. Interstate outbreak of *Salmonella new brunswick* infection traced to powdered milk. *Journal of the American Medical Association*, 1968, 203, 838-44.

12. Rowe B., Hutchinson D.N., Gilbert R.J., Hales B.H., Begg N.T., Dawkins H.C., Jacob M., Rae F.A., Jepson M. *Salmonella ealing* infections associated with consumption of infant dried milk. *Lancet*, 1987, 8564, 900-3.

13. Food and Agriculture Organisation, World Health Organisation. *Enterobacter sakazakii* and *Salmonella* in powdered infant formula: meeting report, Rome, January 2006, in *Microbiological Risk Assessment Series, No. 10*. Ed. Food and Agriculture Organisation, World Health Organisation Geneva, WHO. 2007.

14. International Dairy Federation. Recommendations for the hygienic manufacture of milk and milk based products, in *IDF Bulletin No. 292*. Ed. International Dairy Federation. Brussels, IDF. 1994.

15. Usera M.A., Echeita A.,,Alduena A., Raymundo R., Prieto M. I., Tello O., Cano R., Herrera D., Martinez-Navarro F. Interregional salmonellosis outbreak due to powdered infant formula contaminated with lactose fermenting *Salmonella virchow*. *European Journal of Epidemiology*, 1996, 12 (4), 377-81.

16. Anon. *Salmonella anatum* infection in infants linked to dried milk. *Communicable Disease Report Weekly*, 1997, 7 (5), 33, 36.

17. Anderson P.H.R., Stone D.M. *Staphylococcus* food poisoning associated with spray-dried milk. *Journal of Hygiene*, 1955, 53, 387.

18. Armijo R. Henderson D.A., Timothee R., Robinson H.B. Food Poisoning outbreaks associated with spray-dried milk. An Epidemiologic study. *American Journal of Public Health*, 1957, 47, 1093.

19. EI-Dairouty K.R. Staphylococcal intoxication traced to non-fat dried milk. *Journal of Food Protection*, 1989, 52 (12), 901-2.

20. Asao T., Kumeday Y., Kawai T., Shibata T., Oda H., Haruki K., Nakazawa H., Kozahi S. An extensive outbreak of staphylococcal food poisoning due to low-fat milk in Japan: estimation of enterotoxin A in the incriminated milk and powdered skim milk. *Epidemiology and Infection*, 2003, 130 (1), 33-40.

21. Becker H., Schaller G., von Wiese W., Terplan G. *Bacillus cereus* in infant foods and dried milk products. *International Journal of Food Microbiology*, 1994, 23 (1), 1-15.

22. Pavic S., Brett M., Petric I., Lastre D., Smoljanovic M., Atkinson M., Kovacic A., Cetinic E., Ropac D. An outbreak of food poisoning in a kindergarten caused by milk powder containing toxigenic *Bacillus subtilis* and *Bacillus licheniformis*. *Archiv fur Lebensmittelhygiene*, 2005, 56 (1), 20-2.

23. Holmes J.R., Plunkett T., Pate P., Roper, W., Alexander W.J. Emetic food poisoning caused by *Bacillus cereus*. *Archives of Internal Medicine*, 1981, 141 (6), 766-7.

24. Biering G., Karlsson S., Clark N.C., Jonsdottir K.E., Ludvigsson P., Steingrimsson. Three cases of neonatal meningitis caused by *Enterobacter sakazakii* in powdered milk. *Journal of Clinical Microbiology*, 1989, 27 (9), 2054-6.

25. Nazarowec-White M., Farber J.M. Incidence, survival, and growth of *Enterobacter sakazakii* in infant formula. *Journal of Food Protection*, 1997, 60 (3), 226-30.

26. Raghav M., Aggarwal P.K. Isolation and characterisation of *Enterobacter sakazakii* from milk foods and environment. *Milchwissenschaft*, 2007, 62 (3), 266-9.

27. Arku B., Mullane N., Fox E., Fanning S., Jordan K. *Enterobacter sakazakii* survives spray drying. *International Journal of Dairy Technology*. 2008, 61 (1), 102-10.

28. Shaker R., Osaili T., Al-Omary W., Jaradat Z., Al-Zuby M. Isolation of *Enterobacter sakazakii* and other *Enterobacter* spp. from food and food production environments. *Food Control*, 2007, 18 (10), 1241-5.

29. Galvano F., Galofaro V., Galvano G. Occurrence and stability of aflatoxin M1 in milk and milk products: a worldwide review. *Journal of Food Protection*, 1996, 59 (1 a), 1079-90.

30. Marth E.H. Dairy Products, in *Food and Beverage Mycology*. Ed. Beuchat L.R. New York, Avi Publishers. 1987, 175-209.

3. CREAM

3.1   Definitions

According to the UK Food Labelling Regulations 1996, cream is defined as that part of cows' milk rich in fat that has been separated by skimming or otherwise and which is intended for sale for human consumption.

Cream is often perceived as a luxury item, and therefore purchasing patterns are different from those that apply to milk. For this reason, the required shelf life is longer than for milk, and therefore heat processes are usually greater for cream than for milk.

There are a number of different types of cream, usually classified according to their fat content, and often defined in legislation. Eight types are recognised in the UK.

Minimum fat content (%)

1. Half cream	≥ 12
2. Sterilised half cream	≥ 12
3. Single cream or cream	≥ 18
4. Sterilised (or canned) cream	≥ 23
5. Whipping cream	≥ 35
6. Whipped cream	≥ 35
7. Double cream	≥ 48
8. Clotted cream	≥ 55

- *Whipping or whipped creams* have carefully controlled fat contents, designed to give maximum volume and stability when aerated.

- *Frozen cream* may be produced as an ingredient for further processing, or for retail sale.

- *Clotted cream* is produced by heating a layer of double cream above milk in a shallow tray, followed by slow cooling and removal of the cream. Alternatively, it may be made by scalding, by heating a thin layer of high-fat cream directly in a tray.

- *Cream-based desserts* can be defined as dessert products where milk ingredients make up at least 40% of the dry matter.

All creams are subjected to heat treatment such as pasteurisation or sterilisation. Based on the heat treatment that cream has been subjected to, they are classed as:

- *Untreated cream* that has not been treated by heat or in any manner likely to affect its nature and qualities, and has been derived from milk which has not been so treated.

- *Pasteurised cream* that has been either (a) heated to a temperature not less than 63 °C for 30 minutes, or (b) heated to a temperature not less than 72 °C for 15 seconds, or (c) subjected to equivalent temperature and time, other than those already mentioned, so as to eliminate vegetative pathogenic organisms.

- *Sterilised creams* have been subjected to sterilisation (>108 °C for 45 minutes or the equivalent) by heat treatment in the container. It is supplied to consumers in the same container as that containing the cream during sterilisation.

- *Ultra High Temperature (UHT) creams* are subjected by continuous flow to an appropriate heat treatment (>140 °C for 2 seconds or the equivalent) and are aseptically packaged.

3.2   Initial Microflora

The initial microflora are essentially those of the raw milk (influenced by microflora on the cow's udder, milk-handling equipment, and storage conditions) from which the cream is made.

3.3   Processing and its Effect on the Microflora

The production of cream is outlined in Figure 3.1 on the following page.

3.3.1   Storage and transport of raw milk

Generally, the same comments apply to raw milk for cream production as for fresh milk products, and the milk should be of equivalent microbiological quality. It is important to ensure the milk is produced hygienically, as the heat that cream is subjected to kills vegetative cells but not spores. However, the high fat content of cream means that it is more susceptible to spoilage by extracellular lipases produced by psychrotrophic *Pseudomonas* spp. and other organisms in raw milk. These enzymes can survive heat treatment, and therefore it is preferable to minimise the refrigerated storage time of raw milk for fresh cream production, and process as soon as possible after collection.

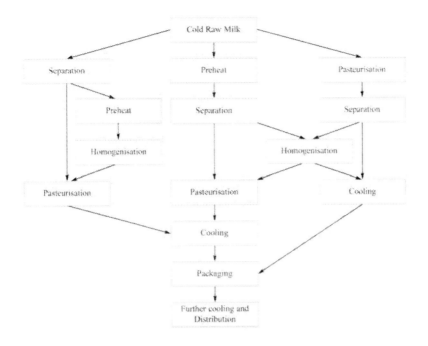

Fig. 3.1. Alternation sequence of operations to produce cream. Reproduced with permission from Wilbey R.A. Microbiology of Cream and Butter, in *Dairy Microbiology Handbook: The Microbiology of Milk and Milk Products*. Ed. Robinson R. New York, John Wiley & Sons, Inc. 2002, 123-74.

3.3.2   Separation

Separation is the concentration of the fat globules and their removal from the milk. Traditionally, this used to be done by skimming, but centrifugal separators are now used in commercial dairies.

Centrifugal separators of the disc stack type are commonly used in modern operations. These consist of a series of conical steel discs within the bowl of the separator, rotated by a spindle. Milk is fed into the rotating bowl and passes into the disc stack through holes. The milk is accelerated and the less dense fat globules move inwards on the disc surface as the heavier serum phase moves outwards. Both phases are then collected in separate chambers. Solid particles of debris and somatic cells in the milk collect on the outer wall of the separator bowl and form a layer of slime. This slime may also contain some bacteria from the milk, particularly clumps or chains of cells. Although it is suggested that separation sometimes concentrates bacteria in the fat phase, however, there seems to be little evidence of a difference in the populations of the two phases (1).

To minimise damage to fat globules, separation is ideally carried out at a temperature of 40 - 50 °C a temperature at which rapid microbial growth is possible. Therefore, higher temperatures (55 – 63 °C) are often recommended; viscous creams are generally produced using these high temperatures. Some separators used to produce high-fat creams (40% fat content) are able to operate at 5 °C, at which temperature significant growth will not occur.

Standardisation of the cream for fat content is usually necessary after separation, since it is difficult to control the process sufficiently to achieve exactly the required level. Separators are therefore set to give a slightly higher than required fat content, and whole or skimmed milk is then added to give the correct value. Standardisation is often carried out at about 40 °C and there is therefore a risk of rapid microbial growth if the process is not carried out quickly. In larger modern dairies this problem can be overcome by partially automating the separation process, either by precise control of flow rates or by feedback control using accurate on-line determination of the fat content in the cream produced.

3.3.3   *Homogenisation*

The need to homogenise cream depends on the particular characteristics of the cream type produced. Half and single creams are usually homogenised to prevent fat separation and provide adequate viscosity. Double and whipping creams are not usually homogenised unless they are UHT-processed.

Homogenisation may be carried out before or after heat treatment, but, from a microbiological point of view, homogenisation before heat treatment is preferred. Homogenisation after heat treatment helps to reduce problems with rancidity caused by lipases present in the milk, and some producers therefore choose this approach. UHT-treated cream is normally homogenised after heat treatment.

3.3.4   *Heat treatment*

Almost all cream sold in developed countries must be heat treated in some way to ensure a safe product.

Minimum pasteurisation treatments are set out in the legislation of many countries and, in the UK, are the same as those applied to pasteurised milk (72 °C for 15 seconds, or 63 - 65 °C for 30 minutes). High-temperature, short time (HTST) processes are almost universally used in modern dairies, but higher temperatures than those used in HTST processing are often applied, both to achieve a longer shelf life, and to overcome the protective effect of the high fat content. For example, the International Dairy Federation (IDF) has recommended a process of 75 °C for 15 seconds for cream with a fat content of 18%, and 80 °C for 15 seconds for cream containing 35% or more fat. In the United States, dairy products containing more than 10% fat should receive a minimum heat treatment of 74.4 °C for 15 seconds.

Cream may also be sterilised in containers either by batch or continuous rotary retorting at 110 - 120 °C for 10 - 20 minutes. For homogenised fat cream, heat treatments of 121 °C for 15 minuntes or 122 °C for 10 minutes are given. Containers must receive a heat treatment of not less than 108 °C for 45 minutes. Cans are sterilised at 116 - 121 °C for 30 minutes, but if the cream receives UHT treatment lower time-temperatures can be used (2). This process is only suitable for cream with a low fat content, since high-fat creams conduct heat poorly.

UHT sterilisation processes are also applied, followed by aseptic filling into cartons in a process similar to that used for milk. A minimum process of 140 °C for 2 seconds is stipulated in the UK to render the cream free of both viable cells and spores, although some very heat-resistant bacterial spores may still survive this process. UHT processing is most suitable for single and half cream. The control of the process becomes increasingly difficult as the fat content rises.

Another method for sterilisation is the Autothermal Thermophilic Aerobic Digester (ATAD) friction process where the milk is initially preheated to 70 °C and subsequently heated to 140 °C for 0.54 seconds. This process can be used for creams containing 12 and 33% fat (2).

3.3.5   Cooling and packaging

Pasteurised cream should be cooled as soon as possible after heat treatment to a temperature of 5 °C or less, to prevent growth of thermoduric organisms, and then be packaged quickly. Most cream for retail sale is now packed in plastic pots sealed with metal foil lids. This type of packaging generally carries very low levels of microbial contamination. However, as with pasteurised milk, the hygienic operation of the filling process is essential to prevent post-pasteurisation contamination. Bulk cream for catering is often packed in 'bag in box' containers, and bulk cream for manufacturing is usually transported in stainless steel tanks, which must be cleaned and sanitised effectively between uses.

3.4   Processing of Other Creams

3.4.1   Whipped cream

Whipped cream contains added sugar and stabilisers. The stabilised cream is then pasteurised and held at 5 °C for 24 hours. Compressed air or nitrogen is then introduced into the mix. This provides an excellent aerobic medium for microbial growth, thereby increasing the chances of spoilage in comparison with liquid cream (2).

3.4.2   Frozen cream

Frozen cream is pasteurised at ≥ 75 °C for 15 seconds. It is then quickly cooled to 1 °C before being frozen in containers, as sheets or pellets, or by direct contact with liquid nitrogen. It is then stored at -18 to -26 °C (2).

3.4.3   Clotted creams

Clotted creams are traditionally made by putting milk in a pan 30 cm in diameter and 20 cm deep where it is held for 12 hours to allow the cream to rise. The pan is then put on a steamer until a layer of solidified cream is formed around the edge. Modern methods involve (a) heating double cream over a layer of skimmed or whole milk, in a large, shallow jacketed tray until a crust is formed or (b) heating a thin layer of high-fat cream (54 – 59 % milk fat) at 77 – 85 °C, to form a crust. The more severe heat treatments result in aerobic spore-formers being the predominant microflora. However, slow cooling and poor hygiene are more likely to lead to the growth of spoilage moulds, coliforms and other contaminants (2).

3.4.4   Cream-based desserts

Cream-based desserts typically undergo heat treatments above pasteurisation, in order to allow cooking of other ingredients such as starch.

3.5   Spoilage

The spoilage of cream is generally similar to that described for liquid milk products. However, because of the difference in purchasing patterns, cream is often required to have a longer shelf life than milk (up to 14 days for pasteurised cream), and containers may be opened and then used by the consumer over several days. The keys to obtaining sufficient shelf life are the microbiological quality of the raw milk, good hygiene in processing, and effective temperature control during distribution and storage.

Cream usually receives more severe heat processes than milk, and the post-heat treatment microbial population therefore consists almost entirely of relatively heat-resistant species. Aerobic spore-forming bacteria survive pasteurisation, and psychrotrophic strains of *Bacillus cereus* may cause 'sweet curdling' and 'bitty cream'. Other, more heat-resistant species, such as *Bacillus licheniformis*, *Bacillus coagulans*, and *Bacillus subtilis*, may survive sterilisation and even UHT processes, and may cause bitterness and thinning in sterilised creams (2). *Bacillus pumilus* and *Bacillus sporothermophilus* are now recognised as potential contaminants in cream, primarily carried over from raw milk. Under UHT conditions, *B. sporothermophilus* has D-values of 3.4 - 7.9 sec and z-values of 13.1 - 14.2 (2). Heat-resistant lipases produced by psychrotrophic bacteria growing in the raw milk may also survive high-temperature processing and cause spoilage in UHT cream.

The keeping quality of cream is greatly affected by the introduction of post-process contamination. Psychrotrophic bacteria such as pseudomonads may contaminate pasteurised cream during processing and are important spoilage organisms. The high fat content of cream means that lipolytic species, such as *Pseudomonas fluorescens* and *Pseudomonas fragi*, are a particular problem. A study of pasteurised double cream showed that pseudomonads were the predominant spoilage organisms (3). Psychrotrophic members of the Enterobacteriaceae are also sometimes involved.

Yeasts and moulds are rarely implicated in the spoilage of cream. Few yeasts are able to ferment lactose, but species such as *Candida lipolyticum* and *Geotrichum candidum* may occasionally spoil bakers' whipping cream where sucrose has been added (4). If, however, other organisms hydrolyse lactose, then the yeast can grow rapidly to produce yeasty or fruity flavours and gas; *Torula cremoris*, *Candida pseudotropicalis* and *Torulopsis sphaerica* have been implicated with such defects (2). Where cream is stored at very low temperatures (0 - 1 °C) to prolong the shelf life, mould growth, usually *Penicillium* spp. may develop on the cream surface (4).

Defective cans or leaking seams could cause spoilage of cream due to entry of bacteria from cooling water or other sources, e.g. a waterborne organism, for example *Proteus*, can cause bitterness and thinning, coliforms can produce gas, and lactococci could result in acid curdling (2).

In the case of cream-based desserts, thermoduric organisms are most likely to be an issue due the more aggressive heat treatments that are used. In addition, the added sugar increases the range of contaminants that could grow in the product. Fruit conserves, if added, will lower the pH of the product thus favouring the growth of yeasts and moulds. With multi-component desserts, both individual components, and blends obtained from their mixing could be responsible for microbial spoilage (2).

3.6   Pathogens: Growth and Survival

In practice, to overcome the protective effect of the higher fat content, cream usually receives a more severe heat treatment than milk. This means that pathogens present in the raw cream are more likely to be destroyed. Unpasteurised cream carries a high risk from the presence of foodborne pathogens, as does raw milk, but the recent safety record of pasteurised cream is good. Although food-poisoning outbreaks have been associated with cream, they are often linked to products filled with, or prepared with cream. In these cases, it is probable that poor hygiene during manufacture, and temperature abuse during storage have been important contributory factors.

3.6.1 Salmonella spp.

Salmonellae will not survive the heat treatment applied to cream, and therefore their presence is likely to be due to post-pasteurisation contamination. The cells are likely to survive for extended periods in contaminated cream, but growth is not possible unless significant temperature abuse occurs. Storage at temperatures below 5 °C will prevent multiplication.

Most of the relatively recent outbreaks recorded have been associated with foods prepared with cream. For example, in 1986 an outbreak of *Salmonella typhimurium* DT40 infection affecting 24 people in the UK was linked to consumption of cream-filled profiteroles (5). A much larger outbreak occurred in Navarra in Spain in 1991, and was reported to have affected approximately 1,000 people. The causative organism was *Salmonella enteritidis*, and the outbreak was associated with the consumption of contaminated confectionery custard and whipped cream (6). In 1992, an outbreak of *S. enteritidis* PT4 infection in Wales was associated with fresh cream cakes, and was found to be a result of contamination of the factory environment by the organism, and inadequate cleaning of the nozzles used to pipe cream into the cakes (7).

More recently, in 1998, an outbreak of *S. typhimurium* DT104 infection affected 86 people in Lancashire. The outbreak was linked to inadequately pasteurised milk from a local dairy, but cream from the same dairy was also recalled (8).

3.6.2 Listeria monocytogenes

There has been some concern that *L. monocytogenes* might be able to survive cream pasteurisation processes and then grow during chilled storage. However, *L. monocytogenes* strain Scott A recorded a D-value of 6 seconds at 68.9 °C in raw cream with a fat content of 38%, indicating that pasteurisation is likely to be effective. The D-value increases to 7.8 sec in inoculated 'sterile' cream. Z-values were calculated as 6.8 °C and 7.1 °C, respectively (9). A later study, using two strains of *L. monocytogenes* suspended in different dairy products, including half and double cream, showed that, although heat resistance did vary, minimum pasteurisation processes would be adequate to eliminate the organism in all products. An investigation into the fate of several strains of *L. monocytogenes* in whipping cream at various storage temperatures recorded generation times of 29 - 46 hours at 4 °C. Populations of approximately 10^7 cells/ml were reached after incubation for 30 days, and, at 8 °C, hazardous levels were reached in only 8 days (10). This indicates that post-pasteurisation contamination of cream could be a potentially serious problem. The same post-process hygiene precautions should be applied for cream as for other high-risk chilled products.

Despite this, although *L. monocytogenes* infection is reported to have been linked to cream on epidemiological evidence (4), such cases have not been confirmed by microbiological investigation.

3.6.3   *Yersinia spp.*

Yersinia enterocolitica is a common contaminant of raw milk, although the majority of the strains isolated are not pathogenic to humans. The organism is heat-sensitive and does not survive pasteurisation, but is capable of psychrotrophic growth. Therefore, it is a potential hazard in cream if introduced as a post-pasteurisation contaminant. A survey of dairy products in Australia recorded an isolation of *Y. enterocolitica* from pasteurised cream (11), but the organism was not detected in cream sampled in the UK over the course of a 3-year survey to determine its incidence in foods (12). There have been no reported outbreaks of *Y. enterocolitica* infection associated with cream.

3.6.4   *Staphylococcus aureus*

Although *Staph. aureus* can often be isolated from raw milk, and is a common cause of mastitis in cows, it does not survive pasteurisation, and cases of staphylococcal food poisoning from pasteurised dairy products are now uncommon. It may be introduced into cream as a post-process contaminant, particularly from infected food handlers. However, it is incapable of growth below about 7 °C, and high numbers will only develop following significant temperature abuse. An investigation of growth and enterotoxin A production by *Staph. aureus* in whey cream showed that growth was limited and that enterotoxin was not produced at detectable levels (13).

Despite this, between 1951 and 1970, six outbreaks of staphylococcal poisoning associated with cream were recorded in England and Wales (14). There have been few recent reports of outbreaks, following significant improvements in hygiene and temperature control. As with *Salmonella*, products prepared or filled with cream are now more likely to be implicated as vehicles of staphylococcal poisoning than cream itself, usually as a result of poor hygiene during handling and temperature abuse.

3.6.5   *Bacillus cereus*

B. cereus is common in milk, and its endospores are able to survive pasteurisation. Some strains are also psychrotrophic, and capable of growth in refrigerated dairy products. Nevertheless, there are very few reports of *B. cereus* food poisoning associated with dairy products. There have been a small number of outbreaks associated with the consumption of pasteurised cream. In 1975 cream found to contain 5×10^6 cfu *B. cereus* caused illness in several people (15). In 1989, two members of the same family became ill after consuming fresh single cream that was later found to contain *B. cereus* at levels of $3 \times 10^7/g$ (16).

3.6.6   *Verotoxigenic Escherichia coli (VTEC)*

VTEC, particularly *Escherichia coli* O157, have been found in raw milk and have caused serious outbreaks of infection associated with consumption of raw or inadequately pasteurised dairy products. An outbreak of *E. coli* O157 infection was recorded in the UK in 1998, associated with consumption of raw cream from a small farm dairy. Seven cases were recorded, with four requiring admission to hospital (17). These organisms are destroyed by properly applied pasteurisation, but if any opportunities for cross-contamination between raw and pasteurised cream exist, recontamination could potentially occur. It is likely that *E. coli* O157 could survive for prolonged periods in cream, but growth in the absence of temperature abuse is improbable. In view of the potentially serious nature of infections caused by VTEC, and the low infective dose, it is important to ensure that such cross-contamination does not occur, since growth may not be required to cause infection.

3.6.7   *Viruses*

Viral hepatitis is the most likely viral infection to be associated with dairy products. In 1975 in Scotland, an outbreak of hepatitis A infection occurred associated with cream consumption. The cause of the outbreak was handling of the cream by an infected cook during preparation (18).

3.7   References

1. Griffiths M.W. Milk and unfermented milk products, in *The Microbiological Safety and Quality of Food, Volume 1*. Eds. Lund B.M., Baird-Parker T.C, Gould G.W. Gaithersburg, Aspen Publishers. 2000, 507-34.

2 Wilbey R.A. Microbiology of Cream and Butter, in *Dairy Microbiology Handbook: The Microbiology of Milk and Milk Products*. Ed. Robinson R. New York, John Wiley & Sons, Inc. 2002, 123-74.

3. Phillips J.D., Griffiths M.W., Muir D.D. Growth and associated enzyme activity of spoilage bacteria in pasteurised double cream. *Journal of the Society of Dairy Technology*, 1981, 34 (3), 113-8.

4. Varnam A.H., Sutherland J.P. Cream and cream-based products, in *Milk and Milk Products: Technology, Chemistry and Microbiology*. Eds. Varnam A.H., Sutherland J.P. London, Chapman and Hall. 1994, 183-223.

5. CDR. Communicable disease associated with milk and dairy products England and Wales 1985-86. *CDR Weekly*, 1987, 49, 3-4.

6. Sesma B., Moreno M., Eguaras J. Foodborne *Salmonella enteritidis* outbreak: A problem of hygiene or technology? An investigation by means of HACCP monitoring, in *Foodborne Infections and Intoxications; Proceedings of the 3rd World Congress, Berlin, June 1992*, Vol.2. Eds. Food and Agriculture Organisation, World Health Organisation. Berlin, Institute of Veterinary Medicine. 1992, 1065-8.

7. Evans M.R., Tromans J.P., Dexter E.L.S., Ribeiro C.D., Gardner D. Consecutive *Salmonella* outbreaks traced to the same bakery. *Epidemiology and Infection*, 1996, 116 (2), 161-7.

8. Anon. Defective pasteurisation linked to outbreak of *Salmonella typhimurium* definitive phage type 104 infection in Lancashire. *CDR Weekly*, 1998, 8 (38), 335, 338.

9. Bradshaw J.G., Peeler J.T., Corwin J.J., Hunt J.M., Twedt R.M. Thermal resistance of *Listeria monocytogenes* in dairy products. *Journal of Food Protection*, 1987, 50 (7), 543-4, 556.

10. Rosenow E.M., Marth E.H. Growth of *Listeria monocytogenes* in skim, whole and chocolate milk, and in whipping cream during incubation at 4, 8, 13, 21 and 35 °C. *Journal of Food Protection*, 1987, 50 (6), 452-9, 63.

11. Hughes D. Isolation of *Yersinia enterocolitica* from milk at a dairy farm in Australia. *Journal of Applied Bacteriology*, 1979, 46 (1), 125-30.

12. Greenwood M.H., Hooper W.L. *Yersinia* spp. in foods and related environments. *Food Microbiology*, 1985, 2 (4), 263-9.

13. Halpin-Dohnalek M.I., Marth E.H. Growth and production of enterotoxin A by *Staphylococcus aureus* in cream. *Journal of Dairy Science*, 1989, 72 (9), 2266-75.

14. Ryser E.T. Public health concerns, in *Applied Dairy Microbiology*. Eds. Marth E.H., Steele J.L. New York, Marcel Dekker. 2001, 397-546.

15. Christiansson A. The toxicology of *Bacillus cereus*, in *Bacillus cereus in Milk and Milk Products*. Ed. International Dairy Federation. Brussels, IDF. 1992, 30-5.

16. Sockett P.N. Communicable disease associated with milk and dairy products: England and Wales 1987-89. *CDR Weekly*, 1991, 1 (Review 1), R9-R12.

17. Anon. Cases of *Escherichia coli* O157 infection associated with unpasteurised cream. *CDR Weekly*, 1998, 8 (43), 377.

18. Chaudhuri A.K.R., Cassie G., Silver M. Outbreak of foodborne type-A hepatitis in greater Glasgow. *Lancet*, 1975, 2 (7927), 223-5.

4. BUTTER AND DAIRY SPREADS

4.1   Definitions

Butter is a water-in-oil emulsion typically consisting of at least 80% fat, 15 - 17% water, and 0.5 - 1% carbohydrate and protein. The two principal types of butter produced are sweet cream butter and ripened cream butter. The UK, Ireland, US, Australia and New Zealand prefer sweet cream butter (pH 6.4 - 6.5), which often contains 1.5 - 2.0% salt. In Europe, cultured (ripened cream), unsalted butter is favoured, in which lactic starter cultures are added to convert the lactose to lactic acid and produce flavour compounds, such as acetoin and diacetyl, from citrate. In many countries, salt and lactic cultures are the only permitted non-dairy additions to butter, although, in the UK and other countries, natural colouring agents, such as annatto, β-carotene and turmeric may be added.

Reduced-fat dairy spreads have a milk fat content of about 50 - 60%. Low-fat dairy spreads contain 39 – 41% fat, and very low-fat spreads have <30% fat. These have a much higher water content than butter or reduced-fat spreads. Where the fat content is below about 20%, these products tend to form a continuous water phase and become oil-in-water emulsions. The high water content results in a much lower level of microbiological stability, and the addition of preservatives, such as potassium sorbate or benzoate, may be permitted. The addition of thickeners, such as gelatin and carbohydrates, may be necessary to maintain the physical stability of the product.

4.2   Initial Microflora

Butter is produced from cream, and the cream is the main source of microorganisms in hygienically produced butter. There is little difference between the microflora of whole milk and that of cream, and therefore any organisms likely to be present in raw milk are also likely to be present in cream, including *Clostridium* spp. and *Bacillus* spp. (1).

4.3 Processing and its Effect on the Microflora

4.3.1 Butter

4.3.1.1 Pasteurisation

An outline process for butter manufacture is shown in Figure 4.1.

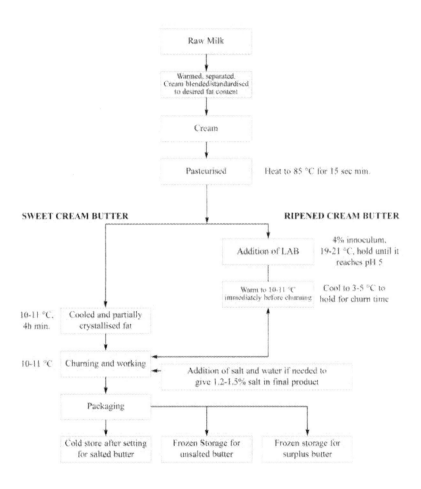

Fig. 4.1. Production of butter. Reproduced with permission from Kornacki J., Flowers R., Bradley R. Jr. Microbiology of Butter and Related Products, in *Applied Dairy Microbiology.* Eds. Marth E., Steele J. New York, Marcel Dekker, Inc. 2001, 127–50.

Cream used for butter production is generally pasteurised after separation, and heat treatments of 85 - 95 °C for 15 - 30 seconds are commonly used. Most vegetative bacterial cells and lactic acid bacteria (LAB) are killed by this process, but bacterial spores, and some thermoduric organisms, such as streptococci and *Microbacterium lacticum* may survive in low numbers (2). Pasteurisation is also important for the destruction of enzymes in the cream, particularly lipases, which might cause hydrolytic rancidity and subsequent flavour defects. Raw cream for butter manufacture should be pasteurised as soon as possible after separation, to minimise the potential for growth of psychrotrophic pseudomonads and other bacteria that produce heat-resistant extracellular enzymes.

4.3.1.2 Cooling

After heat treatment, the cream should be cooled rapidly to 10 – 11 °C and then held for at least 4 hours (1). This allows the completely liquefied butterfat to crystallise into large numbers of small crystals. This process, known as ageing, allows a stable matrix of fat crystals to develop; this is important for the physical properties of the final product. However, if the butter is to be ripened using a starter culture, it is cooled to only 19 - 21 °C. If cooling is too slow, it is possible that bacterial spores surviving pasteurisation could germinate and grow.

4.3.2 Ripened cream butter

Ripened cream butter has traditionally been made by inoculating the pasteurised cream with pure or mixed strains of LAB, then maintaining the temperature at 19 - 21 °C until the required level of acidity is reached (usually 4 - 6 hours). The starter cultures consist of a mixture of acid producers (*Lactococcus lactis* subsp. *lactis* and *Lactococcus lactis* subsp. *cremoris*), and diacetyl-producing species (*Leuconostoc mesenteroides* subsp. *cremoris* and *Lactococcus lactis* biovar *diacetylis*). Cooling the soured butter to 3 – 5 °C stops fermentation; butter is then aged. The reduced pH (about 4.6 - 5) means that the ripened cream and the butter produced from it are more resistant to microbial spoilage than sweet cream butter, but in some countries the soured cream can be neutralised by adding alkaline salts such as sodium carbonate.

An alternative method, the NIZO process, is increasingly being used to produce ripened cream butter. In this process, the cream is treated in the same way as for sweet cream butter (i.e. cooled and aged), but a concentrated diacetyl permeate and lactic starter permeate containing high levels of lactic acid and a flavour-producing starter culture are worked into the butter after the churning stage. Fermentation is allowed to continue for 2 days at 37 °C, and the medium is then ultra-filtered to remove proteins and bacteria. The pH of butter made by this process is more easily adjustable in the desired pH range of 4.8 – 5.3. The resulting butter is said to have improved taste and keeping quality (1, 3).

4.3.2.1 Churning

Most butter is now produced by continuous rather than batch processing, using hygienically designed cleanable-in-place equipment. Churning involves agitation of the cream at low temperature, which produces fat granules that separate from the aqueous phase of the cream to leave buttermilk. The buttermilk is drained off, giving a doubling of the fat content of the cream. Most of the microorganisms present in the cream are retained in the aqueous phase and are therefore removed in the buttermilk. In traditional processes, butter granules are then washed to remove off-flavours, but the wash water itself may be a source of contaminants, such as *Pseudomonas* spp. Modern hygienic processing and continuous production methods do not normally require a washing stage.

4.3.2.2 Salting and working

If the butter is to be salted, the salt is now added to give a concentration of about 2% in the butter. Salt is usually added as a salt-in-water mix, to ensure that the correct water content and salt concentration are maintained. The concentration of salt in the water phase will then be about 11%, sufficient to inhibit the growth of many microorganisms, but this effect is dependent on an even distribution of the salt.

The butter is then worked mechanically both to disperse the salt and water, and to obtain the correct physical structure. The objective is to disperse the water phase into small droplets, preferably between 1- 30 µm in diameter. The microbiological stability of the butter is greatly influenced by this process. If most of the water droplets present are < 10 µm in diameter, any microorganisms within them will not be able to grow and will gradually die off, owing to nutrient depletion and the inhibitory effect of salt or lactic acid (in ripened butter). This effect is termed compartmentalisation. However, if larger water droplets or continuous water channels are present in the butter, as a result of either over- or under-working, the compartmentalisation effect is much reduced, and microbial survival and growth are possible. Inadequate working also leads to formation of moisture droplets, a defect called 'leaky' butter.

4.3.2.3 Packaging

Butter may be packaged either in bulk or in retail size containers. Parchment wrappers are the traditional packaging material, but plastic tubs and laminated foil packs are also common. The packaging should be of good microbiological quality. Bulk butter may be frozen (-30 °C) and stored for periods of up to a year, but good-quality butter stored at chill temperatures generally has an expected shelf life of 6 to 12 weeks. Bulk stored butter may be repackaged prior to sale; this process may cause some redistribution of water droplets, which may affect keeping quality and increase the risk of contamination and subsequent spoilage.

4.3.3   Dairy spreads

4.3.3.1   Emulsification

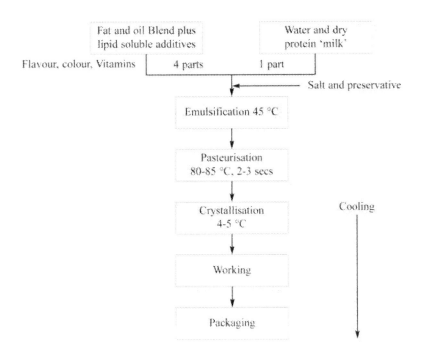

Fig. 4.2. Production of dairy spreads

The process for production of dairy spreads is shown in Figure 4.2. Most spreads are produced on conventional margarine production lines, although high-fat spreads may be produced using a continuous buttermaker. The general method for production of dairy spreads is to add the aqueous phase to the melted fat phase. The lower fat content of reduced- and low-fat spreads means that the composition of the aqueous phase is very important. Salt is added at levels of about 1 - 1.5%, but, in products with relatively high water contents, the final concentration is too low to inhibit microbial growth significantly. Low-fat spreads usually require the addition of thickeners such as gelatin to give the required physical characteristics. Colour, flavours, vitamins and acidulants may also be added. Preservatives, usually sorbic or benzoic acid and their salts, may be added to reduced- and low-fat spreads to improve microbiological stability. The added ingredients should be of good microbiological quality to minimise the initial microbial population. The components are combined in an emulsifying unit at 45 °C, which may be a sufficiently low temperature to allow the growth of thermoduric microbial contaminants such as *Enterococcus faecalis*, *Enterococcus faecium* and thermophiles (1). Since emulsifying units are also difficult to clean effectively, pasteurisation then takes place immediately.

4.3.3.2 Pasteurisation

In-line pasteurisation of the whole emulsion is applied to reduce the level of contaminating microorganisms, which may include potential pathogens. Typically, temperatures of 80 - 85 °C for 2 - 3 seconds are used. This is sufficient to eliminate most vegetative cells, but thermoduric species and bacterial spores may survive. For some reduced- and low-fat products, the aqueous phase is pasteurised before the emulsion is formed. The remainder of the process is then carried out in a closed system to prevent recontamination. The hygienic design and effective cleaning of processing equipment is also of great importance for preventing recontamination of these products (4).

4.3.3.3 Crystallisation and working

Dairy spreads with a fat content of >20% are water-in-oil emulsions. Crystallisation and working are essential to achieve the correct physical properties and texture for the end- product. As with butter, crystallisation is promoted by rapid cooling, which also inhibits the growth of any microorganisms that may have survived pasteurisation. Reduced- and low-fat spreads have much higher water contents than butter, and it is therefore not possible to achieve a small droplet size for the aqueous phase by working. Droplets can be much larger than 10 μm in diameter, and, in low-fat products, continuous water channels are likely. This also has the effect of diluting inhibitors, such as salt or acid, and the compartmentalisation effect described for butter is much reduced. These products are therefore much less resistant to microbial growth and spoilage, and effective hygiene procedures during manufacturing become critical.

4.3.3.4 Packaging

These products may need to be packed in a filtered or sterile air environment to prevent airborne contamination. Very low fat products may also require that the packaging be decontaminated, and the resulting packing process becomes similar to an aseptic filling operation. Such products are usually packed in tubs with a heat-sealed foil laminate lid.

4.4 Spoilage

4.4.1 Butter

4.4.1.1 Bacterial spoilage

Modern hygienic manufacturing methods mean that bacterial spoilage of butter is much less common than in the past. However, defects caused by microorganisms do occasionally occur. Surface taints may develop as a result of growth of *Shewanella putrefaciens* (formerly *Alteromonas putrefaciens*), and *Pseudomonas putrefaciens* or *Flavobacterium* spp. Such spoilage may be apparent within 7 to 10 days of chilled storage. The surface layers are initially affected, but eventually spoilage is apparent throughout the product. A putrid or cheesy flavour develops due to the breakdown of protein (5). Rancidity, proteolytic activity and fruity odours may be caused by the growth of *Pseudomonas fragi* and, occasionally, *Pseudomonas fluorescens* (1). Black discoloration of butter is reported to be caused by *Pseudomonas nigrificans* (1), *Pseudomonas mephitica* is responsible for a skunk-like odour, and an organism formerly known as *Lactococcus lactis* var. *maltigenes* may be responsible for a 'malty' flavour defect linked to the formation of 3-methylbutanal (1, 6). Lipolytic spoilage of butter has been associated with the presence of *Micrococcus* (7).

4.4.1.2 Fungal spoilage

Moulds are still important spoilage organisms for butter, and mould growth may produce surface discolorations and taints. A number of genera have been associated with spoiled butter, including *Penicillium*, *Aspergillus*, *Cladosporium*, *Mucor*, *Geotrichum*, *Alternaria*, and *Rhizopus*. Yeasts may also cause spoilage of butter. Lipolytic species such as *Rhodotorula* may grow on the surface at chill temperatures and may tolerate high salt concentrations. Other yeasts associated with spoilage include *Candida lipolytica*, *Torulopsis*, and *Cryptococcus* (7, 8).

4.4.2 Dairy spreads

There is little information on spoilage of spreads. In theory, the aqueous phase of some low-fat spreads would allow the growth of spoilage bacteria, such as pseudomonads, but in practice the majority of problems are the result of mould growth. Generally, the same genera are involved as for butter spoilage. Preservatives such as sorbic acid help to prevent mould growth, but some species, including *Penicillium* spp. and *Trichoderma harzianum*, are able to convert preservatives to other compounds, which may result in tainting. In low-fat spreads, very low levels of mould contamination may be sufficient to cause spoilage before the end of shelf life (5, 9). The yeast *Yarrowia lipolytica* and bacteria *Bacillus polymyxa* and *E. faecium* have also been reported to be important spoilage organisms in a low-fat dairy spread (1, 8).

4.5   Pathogens: Growth and Survival

4.5.1   Butter

Commercially produced butter is made from pasteurised cream, and that fact, plus its physicochemical characteristics, make it quite inhibitory to bacterial pathogens. It is therefore not surprising that there have been few recorded outbreaks of foodborne disease associated with commercial butter.

4.5.1.1   Staphylococcus aureus

Outbreaks of staphylococcal food poisoning have been associated with butter. In one case, an outbreak involving 24 customers, recorded in the USA in 1970, was linked to whipped butter and to the butter from which the whipped butter was made. The presence of staphylococcal enterotoxin A was demonstrated in both butters. It appeared that the enterotoxin had formed in the cream used to make the butter and was carried over into the finished product (10). A second outbreak, affecting more than 100 customers of pancake houses, was also traced to commercially prepared whipped butter in 1977, and again toxin formation in the cream was suspected (11).

Investigations into the survival of *Staph. aureus* in butter and whipped butter containing 1.5% salt showed that numbers decreased only slowly, especially in whipped butter. Reduction of the salt content to 0 - 1% allowed the population to increase by a factor of ten in 14 days at 23 °C. Therefore a combination of poor hygiene, low salt concentration (or inadequate salt dispersal), and temperature abuse could allow growth of *Staph. aureus* in stored butter (12).

4.5.1.2   Listeria monocytogenes

L. monocytogenes has been shown to grow slowly in butter made from contaminated cream at 4 or 13 °C, and to survive for several months in frozen butter without any appreciable decrease in numbers (13). *Listeria* will not survive cream pasteurisation, but it is a very common environmental contaminant in dairy settings, and effective cleaning and hygiene procedures are necessary to prevent recontamination. Surveys of the incidence of *Listeria* in dairy products have not isolated it from butter (14, 15).

However, despite this, an outbreak of listeriosis associated with butter was reported in a hospital in Finland in 1999. A total of 25 people were affected and six died. A strain of *L. monocytogenes* (serotype 3a) was isolated from packs of butter at the hospital, and from butter and environmental samples at a local dairy plant (16). Butter was also identified as the possible food vehicle in an outbreak of listeriosis, in the US in 1987; 11 pregnancy-associated cases occurred (17). More recently in 2003, 234 cases of listeriosis were reported from 4 clusters in the Humberside and Yorkshire areas of the UK. Environmental samples implicated butter as the cause of the incidence in one cluster (18).

4.5.1.3 Campylobacter

In 1995, an outbreak of *Campylobacter jejuni* enteritis in the USA, which affected 30 people who had eaten in a local restaurant, was associated with garlic butter prepared on site. The survival of *Campylobacter* in butter, with and without garlic, was later investigated, and it was found that *C. jejuni* could survive in butter without garlic for 13 days at 5 °C (19).

4.5.1.4 Toxins

The stability of aflatoxin M1 through butter production and storage has been investigated. Most of the toxin naturally present in the cream was removed with the buttermilk, with very little remaining in the butter. Chilled and frozen storage of the butter had little effect on the toxin (20).

4.5.2 Dairy spreads

There are very few reports of foodborne disease outbreaks associated with dairy spreads, and none associated with reduced- and low-fat products, although it has been suggested that some pathogens may be able to grow in some of these products. Inoculation experiments using two 'light butters' showed that *L. monocytogenes* and *Yersinia enterocolitica* were both capable of growth during refrigerated storage. Both pathogens were capable of more rapid growth than the indigenous microflora (21).

An outbreak of food poisoning caused by *Staphylococcus intermedius* was reported in the USA in 1991. The outbreak affected over 265 people and was associated with consumption of contaminated butter-blend spread (22).

It is likely that pasteurisation and the rigorous hygiene controls applied to the manufacture of these products, especially the low-fat varieties, is effective in preventing the entry of pathogens during processing.

4.6   References

1. Kornacki J., Flowers R., Bradley R. Jr. Microbiology of Butter and Related Products, in *Applied Dairy Microbiology*. Eds. Marth E., Steele J. New York, Marcel Dekker, Inc. 2001, 127–50.

2. The Bacteriology of Butter, in *Dairy Bacteriology*. Eds. Hammer B.W., Babel F.J. New York, Wiley & Sons. 1957.

3. International Dairy Federation. Continuous butter manufacture, in *International Dairy Federation Bulletin 204*. Ed International Dairy Federation. Brussels, International Dairy Federation. 1986, 1-36.

4. Lelieveld H.l.M., Mostert M.A. Hygienic aspects of the design of food plants, in *Food Production, Preservation and Safety*. Ed. Patel P. Chichester, UK, Ellis Horwood Ltd. 1992.

5. Oil- and fat-based foods, in *International Commission on Microbiological Specifications for Foods Microorganisms in Foods, Volume 6: Microbial Ecology of Food Commodities*. Ed. International Commission on Microbiological Specifications for Foods. London, Plenum Publishers. 2005, 480 - 521.

6. Jackson H.W., Morgan M.E. Identity and origin of the malty aroma substance from milk cultures of *Streptococcus lactis* var. *maltigenes*. *Journal of Dairy Science*, 1954, 37, 1316-24.

7. Boor K., Fromm H. Managing microbial spoilage in the dairy industry, in *Food Spoilage Microorganisms*. Ed. Blackburn C. de W. Cambridge, Woodhead Publishing Ltd. 2006, 171-93.

8. Varnam A.H., Sutherland J.P. Butter, margarine and spreads, in *Milk and Milk Products: Technology, Chemistry and Microbiology*. Eds. Varnam A.H., Sutherland J.P. London, Chapman and Hall. 1994, 224-74.

9. Van Zijl M.M., Klapwijk P.M. Yellow fat products (butter, margarine, dairy and nondairy spreads), in *The Microbiological Safety and Quality of Food, Volume 1*. Eds. Lund B.M., Baird-Parker T.C., Gould G.W. Gaithersburg, Aspen Publishers. 2000, 784-806.

10. Anon. Staphylococcal food poisoning traced to butter: Alabama. *Morbidity Mortality. Weekly Report*, 1970, 28, 129-30.

11. Anon. Presumed staphylococcal food poisoning associated with whipped butter. *Morbididity and Mortality Weekly Report*. 1977, 26 (32), 268.

12. Minor T.E., Marth E.H. *Staphylococcus aureus* and enterotoxin A in cream and butter. *Journal of Dairy Science*, 1972, 55 (10), 1410-4.

13. Olsen J.A., Yousef A.E., Marth E.H. Growth and survival of *Listeria monocytogenes* during making and storage of butter. *Milchwissenschaft*, 1988, 43 (8), 487-9.

14. Harvey J., Gilmour A. Occurrence of *Listeria* species in raw milk and dairy products produced in Northern Ireland. *Journal of Applied Microbiology*, 1992, 72 (2), 119-25.

15. Massa S., Cesaroni D., Poda G., Trovatelli L.D. The incidence of *Listeria* spp. in soft cheeses, butter and raw milk in the province of Bologna. *Journal of Applied Bacteriology*, 1990, 68 (2), 153-6.

16. Lyytikainen O., Autio T., Maijala R., Ruutu P., Honkanen-Buzalski T., Miettinen M., Hatakka M., Mikkola J., Anttila V.-J., Johansson T., Rantala L., Aalto T., Korkeala H., Siitonen A. An outbreak of *Listeria monocytogenes* serotype 3a infections from butter in Finland. *Journal of Infectious Diseases*, 2000, 181, 1838-41.

17. Mascola L., Chun L., Thomas J., Bibe W.F., Schwartz B., Salminen C., Heseltine P. A case-control study of a cluster of perinatal listeriosis identified by an active surveillance system in Los Angeles County. *Proceedings of Society for Industrial Microbiology-Comprehensive Conference on Listeria monocytogenes*, Rohnert Park, CA, 1998.

18. CDR. *Listeria monocytogenes* infections in England and Wales in 2004. *Communicable Disease Report Weekly*, 2004, 14 (37).

19. Zhao T., Doyle M.P., Berg D.E. Fate of *Campylobacter* jejuni in butter. *Journal of Food Protection*, 2000, 63 (1), 120-2.

20. Wiseman D.W., Marth E.H. Stability of aflatoxin M 1 during manufacture and storage of a butter-like spread, non-fat dried milk and dried buttermilk. *Journal of Food Protection*, 1983, 46 (7), 633-6.

21. Lanciotti R., Massa S., Guerzoni M.E., Fabio G.D. Light butter: natural microbial population and potential growth of *Listeria monocytogenes* and *Yersinia enterocolitica*. *Letters in Applied Microbiology*, 1992, 15 (6), 256-8.

22. Khambaty F.M., Bennett R.W., Shah D.B. Application of pulsed-field gel electrophoresis to the epidemiological characterisation of *Staphylococcus intermedius* implicated in a food-related outbreak. *Epidemiology and Infection*, 1994, 113 (1), 75-81.

5. CHEESE

5.1 Definitions

Cheese is a stabilised curd of milk solids produced by casein coagulation and entrapment of milk fat in the coagulum. The water content is greatly reduced, in comparison with milk, by the separation and removal of whey from the curd. With the exception of some fresh cheeses, the curd is textured, salted, shaped, and pressed into moulds before storage and curing or ripening.

There are said to be approximately 1,000 named cheeses throughout the world, each produced using a variation on the basic manufacturing process. Most of these varieties fit into one of three main categories according to their moisture content, and method and degree of ripening:

5.1.1 Soft cheese

High moisture (55 - 80%)
a)   fresh, unripened (cottage cheese, Ricotta, Quarg, Fromage Blanc, Neufchâtel, Mozzarella)
b)   surface mould-ripened (Brie, Camembert)

5.1.2 Semi -soft / semi-hard cheese

Moderate moisture (41 - 55%)
a)   surface smear ripened (Limburger, Munster, Tilsit)
b)   ripened by bacteria (Caerphilly, Lancashire, St Paulin)
c)   Blue-veined, internally mould ripened (Stilton, Roquefort, Gorgonzola)

5.1.3 Hard / low moisture cheese

Low moisture (20 - 40%)
a)   ripened by bacteria, with eyes (Emmental, Gruyère)
b)   ripened by bacteria, no eyes (Cheddar, Edam, Cheshire)
c)   very hard (Grana (Parmesan), Asiago, Romano)

5.2   Initial Microflora

Essentially the initial microflora correspond with those of the milk used to produce the cheese.

5.3   Processing and its Effects on Microflora

A diagram of the basic steps in the production of cheese is given in Figure 5.1, using Cheddar as an example.

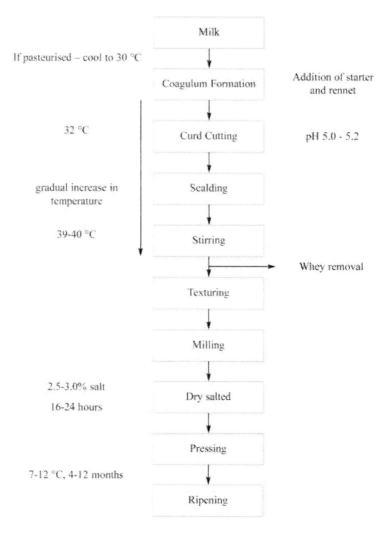

Fig. 5.1. Production of cheese (e.g. Cheddar)

5.3.1 Pasteurisation

Cheese may be made from raw milk, pasteurised milk, or milk that has undergone a sub-pasteurisation (thermisation) treatment. Pasteurisation destroys the vegetative cells of pathogens as well as many spoilage organisms, and some of the enzymes naturally present in the milk. It is argued that pasteurisation affects the ripening and flavour development of cheese, and that only raw milk cheeses develop a full and mature flavour. However, a recent study suggested that, if high quality milk was used, pasteurisation produced differences in texture, but flavour and aroma were little affected (1). A sub-pasteurisation (thermisation) process (typically 65 - 70 °C for 15 - 20 seconds), may be used to destroy many vegetative cells, but without inactivating some of the enzymes involved in flavour development. Milk for cheese may also be subjected to the bactofugation process (see Chapter 1 Liquid Milk Products), which may be used to substantially reduce the number of bacterial spores in the milk, and help to prevent later spoilage.

The principal disadvantage of raw milk is the possible presence of pathogens, such as *Staphylococcus*, *Listeria*, *Salmonella* and verocytotoxigenic *Escherichia coli* (VTEC), all of which have caused outbreaks of infection associated with unpasteurised cheeses. Ideally, from a safety point of view, only pasteurised milk would be used to produce cheese. Despite this, there is a constant demand for unpasteurised cheese, which may be perceived as a superior product. The manufacture of unpasteurised cheeses must be very carefully managed, with the application of effective control measures.

Pasteurised milk for cheese production has a bacterial flora consisting of thermoduric organisms that have survived pasteurisation, such as corynebacteria, micrococci, enterococci, spores of *Bacillus* and *Clostridium*, and post-pasteurisation contaminants, including coliforms and psychrotrophic Gram-negative organisms.

5.3.2 Starter cultures

The acidification of milk is the key step in the making of cheese. Acidification is essential for the development of both flavour and texture; it promotes coagulation; and the reduction in pH inhibits the growth of pathogens and spoilage organisms. It is normally achieved by the fermentation of lactose by bacterial starter cultures to produce lactic acid, although some fresh cheeses, such as cottage cheese, may be acidified by the direct addition of acid, and do not require a starter. In the past, acidification was achieved by the development of the resident microflora of the milk, and this method is still used in some traditional, artisan cheeses. However, this process is difficult to control and tends to give a variable product that may suffer from taints and inconsistent flavours. Therefore, most cheese is now produced using a carefully selected starter, which gives predictable and desirable results. *Lactococcus lactis*, *Streptococcus thermophilus*, *Lactobacillus helveticus* and *Lactobacillus delbrueckii* are the primary species of starter bacteria used in cheese manufacture. The use of frozen, concentrated cultures that can be added directly to the cheese vat is becoming common, for reasons of convenience and to minimise the risk of contamination. For Cheddar, the starter is normally added at a concentration of 10^6 - 10^7 cells/ml.

Some of the commonly used starter organisms used in specific cheeses are shown in Table 5.I. The choice of starter depends on the type of cheese being produced. The temperature of scalding, or cooking, of the curd is an important consideration. Below 30 °C, mesophilic starters, singly, or in combination, are used, such as *L. lactis* subsp. *lactis*, *L. lactis* subsp. *cremoris*, and *Leuconostoc* spp.. Where scalding temperatures are higher (45 - 55 °C), as in Swiss cheeses and very hard cheeses, thermophilic starters (whose optimum growth temperature is 40 °C) are required, such as *Str. thermophilus* and *L. delbrueckii* subsp. *bulgaricus*. Other properties of starter cultures that are important include proteolytic activity, which is important in starter function and flavour development during ripening, and citrate metabolism, which is required for the production of the flavour compound diacetyl in some varieties.

Sometimes, the rate of acid production by the starter is slower than expected. This 'starter failure' can result in a poor quality product and may also enable the growth of pathogens, particularly *Staphylococcus aureus*, before an inhibitory pH is achieved. The most common cause of starter failure is infection of the culture with a bacteriophage. This may be a serious economic problem, but is controlled by careful starter strain selection and the application of rigorous hygiene procedures to prevent contamination. In recent years, there has been much interest in the development of transconjugant starter strains with improved phage resistance. Starter failure may also be caused by the presence of antibiotic residues in the milk, usually as a result of their use to treat animals with mastitis. Therefore, it is normal practice to test all incoming milk for the presence of these residues. Sanitiser residues may also cause starter failure, particularly quaternary ammonium compounds.

Other non-starter microorganisms are also essential for the manufacture of certain types of cheese. For example, *Propionibacterium freudenreichii* is used in the manufacture of some Swiss cheeses, such as Emmental and Gruyère, because it metabolises lactic acid to produce carbon dioxide and propionic acid. The gas is needed for the formation of the characteristic eyes in the cheese, and the propionic acid contributes towards the sweet, nutty flavour of these cheeses. Surface smear-ripened cheeses, such as Brie, Limburger and Munster are ripened using smear-flora that consists of *Brevibacterium linens*, micrococci and yeast. *B. linens* produces an orange-red growth, and is strongly proteolytic contributing to typical odours and flavours in the cheese. Micrococcus spp., for example *Micrococcus virans*, *Micrococcus caselyticus* and *Micrococcus freudenreichii*, promote proteolysis during ripening and are responsible for the characteristic yellow to deep red colour of the cheese surface. Yeasts e.g. *Geotrichum candidum*, *Candida* spp. and *Debaryomyces* spp. contribute to flavour and colour development. Soft cheeses such as Brie and Camembert are ripened by the surface growth of mould spores (*Penicillium camembertii* and (or) *Penicillium caseicolum*), and blue-veined cheeses such as Stilton and Roquefort rely on inoculation of the body of the cheese with *Penicillum roquefortii* spores, plus aeration, to ripen.

Recently there has been some interest in the addition of probiotic organisms to cheese, which are claimed to improve gastrointestinal health, to cheese. Probiotic strains of *Lactobacillus acidophilus* and *Lactobacillus rhamnosus* have been added to fresh cheeses, but most strains do not survive the ripening process in other varieties.

5.3.3   Curd formation

In curd cheeses, a coagulant is normally added to the acidified milk. For varieties such as Cheddar, this is done approximately 30 - 45 minutes after adding the starter, but in other cheeses acidification may be allowed to proceed further.

TABLE 5.I
Lactic acid bacteria (LAB) employed as starter cultures

Bacteria	Examples of usage
L. lactis subsp. *cremoris*, *L. lactis* subsp. *lactis* *Leuconostoc* spp.	Soft, unripened cheese e.g. Cottage Quarg, cream cheese, Neufchâtel
Str. thermophilus, *L. delbrueckii* subsp. *bulgaricus*, *L. helveticus*	Soft, unripened cheese (rennet-coagulated) e.g. Mozzarella
L. lactis spp. *cremoris* *P. camembertii*, Yeasts	Surface mould-ripened cheese e.g. Brie, Camembert, Coulommier
L. lactis spp. *cremoris*, *L. lactis* subsp. *lactis*	Surface bacterial smear-ripened cheese e.g. Limburger
Str. thermophilus, *L. delbrueckii* subsp. *bulgaricus*	Pickled cheese e.g. Feta
L. lactis subsp. *cremoris* and *L. lactis* subsp. *lactis* or *L. lactis* subsp. *cremoris* alone	Hard-pressed cheese e.g. Cheddar, Cheshire, Dunlop, Derby, Double Gloucester, Leicester Semi-hard cheese e.g. Gouda, Edam Lancashire, Caerphilly
Str. thermophilus with *L. delbrueckii* subsp. *lactis* or *bulgaricus*	Hard cheese with eyes e.g. Emmental *Gruy*ère Very hard cheese e.g. Parmesan, Asiago
L. lactis subsp. *lactis* *Lactococcus lactis* biovar *diacetylactis*	Blue-veined cheese e.g. Stilton, Danish Blue, Roquefort, Mycelia, Gammelost

Traditionally, enzymic coagulation by rennet, made from the stomachs of young calves, was used. Recently, however, concerns about shortages of animal rennet, and increasing demand for vegetarian cheeses, have generated interest in microbial rennet. This may consist of acid proteinases produced by moulds such as *Mucor miehei*, or chymosin (the most important component of rennet) produced by fermentation using genetically modified bacteria.

Rennet, in combination with acid from the starter, causes coagulation of the milk curd by precipitating casein as an aqueous gel. The curd is then allowed to set for a time depending on the cheese variety. For most hard or semi-hard cheeses, this would be approximately one hour. During this time, the curd becomes more rigid and its water-retaining capacity decreases for Cheddar, an acidity of about 0.1 - 0.2% is reached, at which point the curd is cut.

Cutting the curd into small cubes leads to syneresis (expulsion of whey and contraction of curd). The mixture is then scalded or cooked at a temperature determined by the cheese variety (38 - 40 °C for Cheddar). This process helps to expel more whey and is important in producing the correct curd characteristics.

When the acidity and curd firmness reach the correct level, the whey is separated from the curd. In Cheddar-type cheeses, the curd is then subjected to a process of compressing and stretching (cheddaring), which fuses the curd into a mat. Traditionally, this was done manually, by piling and turning slabs of curd, but the process is now mechanised in cheddaring towers. The starter bacteria continue to grow during this process, reaching a population of 10^8 - 10^9 cells /ml, and a final acidity of 0.6 - 0.7%. The curd is then milled, salted, moulded and pressed.

Throughout this process, it should be noted that the temperature is maintained at a suitable level for starter growth. This temperature will also favour the growth of contaminating spoilage organisms.

5.3.4   Salting/brining

In the manufacture of Cheddar, salt is added to the milled curd before pressing (dry salting) at a concentration of 1.5 - 2% w/w. In other varieties, such as Gouda and Camembert, the moulded cheese is immersed in a concentrated brine. Some blue cheeses are salted by rubbing dry salt into the surface of the moulded cheese. Salting inhibits the growth of the starter culture and other microorganisms, contributes to the flavour, and affects texture.

5.3.5   Ripening

All but fresh cheeses require some degree of ripening for the full development of flavour and texture. During ripening, further moisture loss occurs, and a complex combination of microbial and enzymic reactions take place, involving milk enzymes, the coagulant, and proteases and peptidases from the starter culture and non-starter organisms, which remain viable although their growth is inhibited. Ripening conditions vary with cheese variety. Soft, high-moisture cheeses are ripened for relatively short periods, whereas hard, strongly flavoured cheeses may ripen for more than a year. Surface-ripened cheeses require quite a high humidity, but most hard cheeses must be kept in dry conditions to inhibit surface microbial growth. Temperature also varies, and Cheddar is normally ripened at approximately 10 °C. Blue-veined cheeses are made to have an open texture so that sufficient oxygen is present in the cheese to allow the growth of *P. roquefortii* throughout, but the process may be assisted by piercing the cheese with metal rods to improve gaseous exchange.

5.4   Processed Cheese

Processed cheeses are produced by milling and mixing naturally-produced cheeses until a plastic mass is formed, usually with additional ingredients such as cream, dry milk, whey, and emulsifying salts such as polyphosphates. The mass is then melted and heated at temperatures of 85 - 95 °C, or as high as 115 °C for several minutes. The molten cheese is then formed into slices or portions and packaged. Some products are processed at a sufficiently high temperature to render them ambient-stable if sufficient preservatives such as salt, lactic acid and potassium sorbate are present.

5.5   Value-added Cheese

Traditional cheese varieties are increasingly modified to create new products by the addition of ingredients such as herbs, nuts and dried fruits. Different varieties may also be processed and then combined to form layered products. The microbiology of these products can be complex, since both the microflora and environmental conditions are altered by the addition of new ingredients. The safety and stability of these cheeses must be carefully considered during development.

5.6   Spoilage

Microbial spoilage of cheese can be caused by both bacteria and fungi, but the type of spoilage depends very much on the characteristics of individual cheese varieties. Both visual and organoleptic defects may result, either on the surface of the cheese or internally.

5.6.1   Fungal spoilage

Although the growth of moulds on the surface or in the body of some cheese varieties is essential for ripening, mould growth is generally not desirable. Mould spoilage is usually unpleasant in appearance, and may result in musty taints and odours. Moulds are also responsible for liquefaction of the curd. There is also the possibility of mycotoxin production in some cases. Moulds commonly involved in cheese spoilage include members of the genera *Penicillium, Aspergillus, Cladosporium, Mucor, Fusarium, Monilia* and *Alternaria* (2). Effective hygiene is important in the control of mould spoilage in cheese, particularly in ripening rooms, and rigorous cleaning procedures are needed to prevent the accumulation of mould spores. Filtered sterile air supplies, or ultraviolet light treatment may also be used to control contamination. The use of vacuum and modified-atmosphere packaging helps to prevent mould growth on pre-packed cheese, but growth may still occur in residual air pockets or in packs that are improperly sealed or become punctured. Where permitted, antifungal agents such as sorbic acid or natamycin may be incorporated into packaging.

Yeasts may cause spoilage of fresh cheeses, such as cottage cheese, during storage, resulting in gas production and off-flavours and odours. Yeast may also proliferate on the surface of ripened cheeses, especially if the surface becomes wet, causing slime formation. Yeasts most frequently isolated from spoiled cheese include *Candida* spp., *Yarrowia lipolytica*, *Pichia* spp., *Kluyveromyces marxianus*, *G. candidum* and *Debaryomyces hansenii* (2 - 5).

5.6.2 *Bacterial spoilage*

In fresh cheeses with a sufficiently high pH, such as cottage cheese, bacterial spoilage may occur. This is likely to be caused by Gram-negative, psychrotrophic species, such as pseudomonads and some coliforms. These organisms may contaminate the product through water used to wash the curd.

Pseudomonas spp., *Alcaligenes* spp., *Achromobacter* spp. and *Flavobacterium* spp. are the psychrotrophic bacteria of concern. *Pseudomonas fluorescens, Pseudomonas fragi* and *Pseudomonas putida* cause bitterness, putrefaction and a rancid odour, liquefaction, gelatinisation of curd, and slime and mucous formation on cheese surfaces. *Alcaligenes viscolactis* is responsible for ropiness and sliminess in cottage cheese, and *Alcaligenes metacaligenes* for 'flat, flavourlessness' in cottage cheese. Psychrotropic *Bacillus* spp. cause bitterness and proteolytic defects (6).

Bacteria may also cause spoilage by the production of internal gas in the cheese, resulting in slits, small holes or blown packs. This may happen in fresh cheese, early in the ripening phase ('early blowing'), or well into the ripening stage ('late blowing'). Early blowing is usually caused by members of the Enterobacteriaceae, but other organisms, such as *Bacillus* spp., are sometimes involved. The problem can be effectively controlled by adequate hygiene and process control in manufacturing.

Late blowing, which may occur after 10 days in varieties such as Gouda, or after several months in some Swiss cheeses, is caused by clostridia that are able to produce butyric acid from lactate. Late blowing sometimes also occurs in Cheddar. Species commonly involved are *Clostridium butyricum, Clostridium tyrobutyricum* and *Clostridium sporogenes*, spores of which survive pasteurisation and can be present in cheese milk. Contamination of milk with these organisms is often seasonal (*C. tyrobutyricum* is more prevalent in winter), and is related to the inclusion of silage in the diet of dairy cows. A very low level of contamination may be sufficient to cause late blowing. In some countries, nisin, a natural antimicrobial produced by strains of *L. lactis*, has been used successfully to control late blowing, by inhibiting the growth of clostridia.

Small, irregular slits may also sometimes appear in 3- to 6-week-old Cheddar, and this 'intermediate blowing' is thought to be associated with the presence of non-starter gas-producing lactobacilli.

5.6.3   Discolouration

Yeast and enterococci have been found responsible for white spots on brine-salted cheeses (2). Surface mould growth by species such as *Aspergillus niger*, may cause discoloration of hard cheeses. Discoloration within the cheese is not common, but pigmented strains of certain lactobacilli have been linked with 'rusty spots' in some cheeses, and non-starter *Propionibacterium* spp. may cause brown or red spots in Swiss cheese (2, 7).

P. fluorescens forms water-soluble pigments while other pseudomonads cause darkening and yellowing of curd. Yellow discolouration may be attributed to flavin pigment formation by *Flavobacterium* spp., and *Bacillus* spp. have been associated with dark pigment formation (6).

5.7   Pathogens: Growth and Survival

The safety record for cheese is relatively good considering the very large quantities that are consumed worldwide. However, there have been a number of serious outbreaks of foodborne disease associated with cheese, and these are well documented. The most serious outbreaks have been caused by *Listeria monocytogenes*, salmonellae and enteropathogenic *Escherichia coli* (EPEC). In recent years, a number of *E.coli* O157 outbreaks, linked to cheese, have been recorded. Cheeses made from raw milk are particularly at risk since they may become contaminated by pathogens initially present in the milk. Pathogens may also enter cheese during processing, if hygiene and process controls are inadequate.

The characteristics of individual cheese varieties greatly influence the potential presence and survival of pathogens. Process and storage temperature, acid production by starter cultures and the addition of salt are all important. In general, soft and semi-soft cheeses with high water activities present fewer barriers to pathogen survival and growth than do hard cheeses. For example *Listeria* is able to multiply in soft, surface ripened cheeses, such as Brie and Camembert, but is unable to grow in properly made Cheddar, although it may survive for long periods.

5.7.1   Listeria spp.

An outbreak of listeriosis in California, in 1985 involved 142 cases and resulted in 48 deaths. This outbreak was associated with Mexican-style cheese, and contributed greatly to the establishment of L. monocytogenes as a foodborne pathogen (9). The processing environment and equipment were found to be contaminated and the proper pasteurisation of the cheese milk was questioned (10 - 12). During the period 1983 - 1987 other serious outbreaks associated with soft cheeses, such as Vacherin Mont d'Or, in Switzerland, were reported. In 1995, an outbreak in France causing four deaths was linked to Brie de Meaux cheese made from raw milk (13), and in early 2000 a further outbreak in France was linked to soft cheese. The sale of illicitly produced or distributed fresh, unripened cheese made from raw milk has been associated with several listerial outbreaks as recently as 2003, in the Hispanic community. This outbreak resulted in one foetal death and the death of a neonate (14). These outbreaks illustrate the serious problems posed by L. monocytogenes in cheese production.

L. monocytogenes is a psychrotrophic, fairly heat-tolerant pathogen, ubiquitous in the environment, and can also be found in raw milk. It may therefore enter the cheese process by a variety of routes, particularly in smaller, traditional operations where hygiene procedures may be poor. Surface-ripened cheeses are especially vulnerable to recontamination and growth of the organism. As the ripening process proceeds, the development of mould on the surface raises the pH from around 5.0 up to 6.0 - 7.0. This, combined with the high moisture content and temperature of the ripening rooms (8 - 12 °C), creates conditions in which rapid growth of L. monocytogenes is possible. Counts of 10^7 cfu/g have been demonstrated at the surface of Camembert after 56 days (15). The same process may occur during the ripening of blue-veined cheeses.

Although growth of Listeria is much less likely to occur in other cheese varieties where there is no rise in pH during ripening, the organism may survive for long periods. For example, viable cells have been found in Cheddar cheese stored for 434 days (16), and raw-milk soft or semi-hard cheese that had undergone aging for approximately 60 days was implicated in an outbreak in Canada in 2002 (14). This casts some doubt on the recommendation to hold Cheddar and some other hard cheeses made from raw milk, at, or above, 1.7 °C for at least 60 days as a control for Listeria and other pathogens.

For these reasons, it is essential that adequate hygiene procedures are practised during cheese manufacture and ripening to prevent environmental contamination with L. monocytogenes. Environmental testing for the organism is also recommended. This is equally true for cheese made from raw or pasteurised milk. In addition, control of the bacteriological quality of raw milk used to make cheese is important, and can help to reduce the incidence of Listeria in raw milk cheeses. End product testing is also widely practised with susceptible cheese varieties, but this can never be sufficient to assure the safety of the product.

Surface-ripened soft cheeses made from raw milk are inherently hazardous products, although the amount of attention given to this problem has led to recent improvements. In 1996, a UK survey of raw milk soft cheeses showed only one of 72 samples tested contained *L. monocytogenes* (21), at a level of <10 cfu/g. Also in 1996, a survey of *Listeria* in acid curd cheese conducted in Germany showed that 9 of 50 samples were positive for *Listeria*; 2 of which were *L. monocytogenes* also at a level of less than 10 cfu/g (38). Despite this, vulnerable consumers in the UK, such as pregnant women and the elderly, are advised not to eat surface-ripened soft cheese in general.

5.7.2   *Escherichia coli*

EPEC is a cause of gastroenteritis in humans, but its growth is usually inhibited in cheese manufacture by acidity and pH. However, if starter activity is impaired, *E.coli* levels may reach high levels during cheese-making and may survive in the finished product. In the USA in 1971, EPEC was the cause of a major outbreak associated with imported Camembert, in which 387 cases were involved (19). This outbreak was found to be the result of post-pasteurisation contamination.

Enterohaemorrhagic strains of *Escherichia coli* (EHEC), such as *E.coli* O157:H7, are important emerging pathogens and may cause serious infections and fatalities in the very young and the elderly. *E.coli* O157:H7 is considered to be a potentially high-risk pathogen in cheese, because of its unusual ability to tolerate low pH values for long periods and its association with unpasteurised milk. It has been shown to survive the cheese-making process (20), and has been found in surveys analysing cheese (21, 22). Considering the relatively low infective dose associated with *E.coli* O157:H7 infection, it is perhaps surprising that there have been very few outbreaks associated with cheese. However, several outbreaks have occurred, including one associated with unpasteurised soft cheese in France, and a small outbreak in the north-east of England in early 1999, associated with unpasteurised Cotherstone cheese (23). In the most recent survey of *E.coli* in unpasteurised cheese in the UK in 1999, *E.coli* O157:H7 was not found in any of 801 samples (24).

It would be expected that surface ripened soft cheeses made from raw milk would be particularly high-risk products for infection with these organisms, but as yet there have been no reported outbreaks associated with them. Nevertheless, if raw milk is used in soft cheese manufacture, it is difficult or even impossible to assure the safety of the product in this respect. Control of raw milk quality and checks on the health of dairy cattle are essential for unpasteurised soft cheeses, but it should be noted that *E.coli* O157:H7 is not notably heat resistant, and is effectively killed by standard pasteurisation processes.

5.7.3   *Salmonella spp.*

Salmonellae can be isolated in milk from infected animals, but do not survive pasteurisation. However, salmonellae are quite resistant to other environmental factors, and, in cheese made from raw milk, they may survive through the cheese-making process and be present in the finished product. They may also enter cheese as post-pasteurisation contaminants if the process is not properly controlled. If acid production during manufacture is slow, salmonellae may grow during cheese-making (25), and have been shown to survive for more than 60 days in some cheeses (26, 27).

There have been a number of large outbreaks of salmonellosis associated with cheese. For example, in 1976 in the US, pasteurised Cheddar cheese contaminated with *Salmonella heidelberg* was linked to an outbreak of 28,000 - 36,000 cases of illness (28). In a further outbreak in 1982, in Canada, unpasteurised Cheddar cheese contaminated with *Salmonella muenster* was implicated, and in this case the organism could still be detected after 125 days' storage (29). Another Canadian outbreak, involving Cheddar contaminated with *Salmonella typhimurium*, occurred in 1984, and over 1,500 cases were confirmed. Both these outbreaks in Canada serve to illustrate the ability of salmonellae to survive for long periods in cheese at refrigeration temperatures (26). More recently, in 1996, an outbreak of *Salmonella gold-coast* infection in the UK involved over 80 cases, and was associated with a Cheddar-type cheese (30). There have also been outbreaks linked to other cheese varieties, including Mozzarella.

5.7.4 *Staphylococcus aureus*

In 1999, one case of enterotoxin food-poisoning was reported in Brazil. Minas cheese was implicated in the outbreak involving 50 people. The food handlers were found to be the source of contamination (31).

Low numbers of *Staph. aureus* are relatively common in raw milk. This may due to contamination from the udder surface during milking, and/or shedding of the organism into milk by cattle with sub-clinical staphylococcal mastitis. Before the mid 1960s it was a common cause of cheese-associated food poisoning, for several reasons. *Staph. aureus* is able to tolerate salt and moderate acidity, and can multiply during cheese manufacture and ripening in soft cheeses. It may survive for long periods even in hard cheeses, and, if high enough populations are developed, enterotoxin may be produced, which persists for many months even after the cells have lost viability (32).

The hazard can easily be controlled by storage of raw milk at temperatures that prevent staphylococcal growth, followed by pasteurisation and adequate post-process hygiene to prevent recontamination. There have been few outbreaks in the last 30 years, and these have been mostly associated with slow acid production by starters, or faulty process control, allowing populations of $>10^6$ cfu/g to develop in cheese (10-12). At these levels, toxin production is possible if the acidity is higher than normal. An abnormally high cheese pH value at milling should result in analysis for *Staph. aureus* and then the presence of enterotoxins. Viable cell numbers decrease rapidly during ripening, but analysis for thermostable nuclease (TNA) can provide an indication of staphylococcal growth (33, 34). All samples giving a positive TNA result should be tested for enterotoxin.

5.7.5   Clostridium botulinum

Although botulism is rarely associated with dairy products, there have been a few outbreaks of the disease associated with cheese. Several cases of type B botulism identified in France and Switzerland in the mid 1970s were linked with Brie. The cheese was found to have been ripened on straw beds, which were the probable source of the organism (35). There have also been two outbreaks associated with processed cheese spreads. These spreads contained up to 60% moisture and had pH values of 5.2, conditions suitable for growth and toxin production (36). More recently, in Italy in 1996, an outbreak was linked to Mascarpone that had been subjected to temperature abuse; it resulted in 8 cases of food poisoning with one death (37, 38).

5.7.6   Other pathogens

Outbreaks of infection with other foodborne pathogens have rarely been associated with cheese. There are recorded outbreaks of gastroenteritis caused by *Shigella* spp. contamination in cheese, but no recorded outbreaks caused by *Campylobacter*, *Clostridium perfringens* or *Bacillus cereus* (32).

Both *Brucella abortus* and *Brucella melitensis* can be found in raw milk produced in areas where brucellosis is still present. Both species may survive the cheese-making process and may survive for several months, even in hard cheeses. Small outbreaks of human brucellosis have been associated with home-made raw milk cheeses produced from contaminated milk (10 - 12).

Some viruses that are pathogenic to humans, such as Hepatitis A, are known to contaminate milk, but little is known about their ability to survive in cheese, and their potential for infection. However, poliovirus has been shown to survive in unpasteurised Cheddar for 7 months (39).

5.7.7   Toxins

Mycotoxins could be present in cheese either as a result of the use of contaminated milk, or as a consequence of mould growth on the cheese during ripening and storage. Mycotoxins in milk are usually due to the use of mould-contaminated animal feed, which should be avoided, since aflatoxins in particular are quite stable and are likely to persist through manufacture and ripening. Mycotoxins may also be produced directly in cheeses as a result of the growth of contaminating organisms on the surface, or by inoculated cultures used in mould-ripened cheeses. For example, *P. camembertii* has been shown to produce cyclopiazonic acid, and *P. roquefortii* has also been reported to produce mycotoxins (40). However, the ability to produce toxins does not necessarily mean that production will occur in cheese. Indeed, mycotoxins have rarely been found in cheese, and are generally isolated only when extensive and unacceptable mould growth is apparent.

Two biogenic amines, tyramine and histamine, may be found in cheese. They are produced as a result of the decarboxylation of the amino acids, tyrosine and histidine, by LAB and certain Enterobacteriaceae during cheese ripening. Both amines are vasoactive at high levels. Tyramine can cause a significant, and serious, increase in blood pressure, and histamine has the opposite, hypotensive, effect. Histamine poisoning can result in a variety of symptoms, including flushing, headaches, nausea and a rise in pulse rate. Low levels of biogenic amines are not normally hazardous, since they are deaminated by enzymes in the gut, although individuals deficient in these enzymes may be at risk. Cases of both tyramine-induced hypertensive crisis, and histamine poisoning, have been reported to be associated with cheese (32). Only ripened cheeses contain sufficient precursors, histidine and tyrosine, for significant build-up of amines to occur, and levels sufficient to cause illness are most often associated with mould-ripened cheeses and cheese made from raw milk.

5.8   References

1. Nicol K., Robinson R. The taste test. *Dairy Industries International*, 1999, 64 (8), 35, 37, 38.

2. Johnson M.E. Cheese products, in *Applied Dairy Microbiology*. Eds. Marth E.H., Steele J.L. New York, Marcel Dekker. 2002, 345-384.

3. Fleet G.H. Yeasts in dairy products. A review. *Journal of Applied Bacteriology*, 1990, 68 (3), 199-211.

4. Rohm H., Eliskases-Lechner F., Brauer M. Diversity of yeasts in selected dairy products. *Journal of Applied Bacteriology*, 1992, 72 (5), 370-6.

5. Viljoen B.C., Greyling T. Yeasts associated with Cheddar and Gouda making. *International Journal of Food Microbiology*, 1995, 28 (1), 79-88.

6. Ed. Robinson R.K. *Dairy Microbiology Handbook: The Microbiology of Milk and Milk Products*. New York, John Wiley & Sons. 2002.

7. Baer A., Ryba I. Serological identification of propionibacteria in milk and cheese samples. *International Dairy Journal*, 1992, 2 (5), 299-310.

8. Britz T.J., Riedel K.-H.J. Propionibacterium species diversity in Leerdammer cheese. *International Journal of Food Microbiology*, 1994, 22 (4), 257-67.

9. Linnan M.J., Mascola l., Lou X.D. Epidemic listeriosis associated with Mexican-style cheese. *New England Journal of Medicine*, 1988, 319 (13), 823-8.

10. Johnson E.A., Nelson J.H., Johnson M. Microbiological safety of cheese made from heat-treated milk, Part 1. Executive summary, introduction and history. *Journal of Food Protection*, 1990, 53 (5), 441-52.

11. Johnson E.A., Nelson J.H., Johnson M. Microbiological safety of cheese made from heat-treated milk, Part 11. Microbiology. *Journal of Food Protection*, 1990, 53 (6), 519-40.

12. Johnson E.A., Nelson J.H., Johnson M. Microbiological safety of cheese made from heat-treated milk, Part Ill. Technology, discussion, recommendations, bibliography. *Journal of Food Protection*, 1990, 53 (8), 610-23.

13. Goulet V., Jaquet C., Vaillant V., Rebiere I., Mouret E., Lorente C., Maillot E., Stainer F., Rocourt J. Listeriosis from consumption of raw-milk cheese. *Lancet*, 1995, 345 (8964), 1581-2.

14. Norton D.M., Braden C.R. Foodborne listeriosis, in *Listeria, Listeriosis and Food Safety*. Eds. Ryser E.T., Marth E.H. New York, Marcel Dekker. 2007, 305-56.

15. Ryser E.T., Marth E.H. Fate of *Listeria monocytogenes* during the manufacture and ripening of Camembert cheese. *Journal of Food Protection*, 1987, 50 (5), 372-8.

16. Ryser L.T., Marth E.H. Behavior of *Listeria monocytogenes* during the manufacture and ripening of Cheddar cheese. *Journal of Food Protection*, 1987, 50 (1), 7-13.

17. Nichols G., Greenwood M., de Louvois J. The microbiological quality of soft cheese. *PHLS Microbiology Digest*, 1996, 13 (2), 68-75.

18. Rudolf M., Scherer S. Incidence of *Listeria* and *Listeria monocytogenes* in acid curd cheese. *Archiv für Lebensmittelhygiene*, 2000, 51 (4 - 5), 118-20.

19. Marrier, R., Wells, J.G., Swanson, R.C, Callahan, W., Mehlman, I.J. An outbreak of enteropathogenic *Escherichia coli* foodborne disease traced to imported French cheese. *Lancet*, 1973, 2, 1376-8.

20. Reitsema C.J., Henning D.R. Survival of enterohemorrhagic *Escherichia coli* O157:H7 during the manufacture and curing of cheese. *Journal of Food Protection*, 1996, 59 (5), 460-4.

21. Knappstein K., Hahn G., Heeschen W. Studies on the occurrence of verotoxin producing *Escherichia coli* in soft cheese. *Archiv für Lebensmittelhygiene*, 1996, 47 (3), 59-62.

22. Quinto E.J., Cepeda A. Incidence of toxigenic *Escherichia coli* in soft cheese made with raw or pasteurised milk. *Letters in Applied Microbiology*, 1997, 24 (4), 291-5.

23. Anon. *Escherichia coli* O157 associated with eating unpasteurised cheese. *Communicable Disease Report Weekly*, 1999, 9 (13) 113-6.

24. Ministry of Agriculture, Fisheries and Food, Joint Food Safety and Standards Group London. Report on a study of *E.coli* in unpasteurised milk cheeses on retail sale. MAFF. 2000.

25. Hargrove R.E., McDonough F.E., & Mattingly J.A. Factors affecting survival of *Salmonella* in Cheddar and Colby cheese. *Journal of Milk and Food Technology*, 1969, 32, 480-4.

26. D'Aoust J.-Y., Warburton D.W., Sewell A.M. *Salmonella typhimurium* phage-type 10 from Cheddar cheese implicated in a major Canadian food borne outbreak. *Journal of Food Protection*, 1985, 48 (12), 1062-6.

27. White C.H., Custer E.W. Survival of *Salmonella* in Cheddar cheese. *Journal of Milk and Food Technology*, 1976, 39 (5), 328-31.

28. Fontaine R.E., Cohen M.l., Martin W.T., Vernon T.M. Epidemic salmonellosis from Cheddar cheese: Surveillance and prevention. *American Journal of Epidemiology*, 1980, 11 (2), 247-53.

29. Wood D.S., Collins-Thompson D.L., Irvine D.M., Myhr A.N. Source and persistence of *Salmonella muenster* in naturally contaminated Cheddar cheese. *Journal of Food Protection*, 1984, 47 (1), 20-2.

30. Anon. *Salmonella gold-coast. Communicable Disease Report Weekly*, 1996, 6 (51), 443.

31. Simeão do Carmoa L., Souza Diasb R., Roberto Linardic V., José de Senad M., Aparecida dos Santose D., Eduardo de Fariaf M., Castro Penaf E., Jettg M., Guilherme Heneineh L.. Food poisoning due to enterotoxigenic strains of *Staphylococcus* present in Minas cheese and raw milk in Brazil. *Food Microbiology*, 2002 , 19 (1), 9-14.

32. Teuber M. Fermented milk products, in *The Microbiological Safety and Quality of Food, Volume 1*. Eds. Lund B.M., Baird-Parker T.C., Gould G.W. Gaithersburg, Aspen Publishers. 2000, 535-89.

33. Park C.E., El Derea H.B., Rayman M.K. Evaluation of staphylococcal thermonuclease (TNase) assay as a means of screening foods for growth of staphylococci and possible enterotoxin production. *Canadian Journal of Microbiology*, 1978, 24 (10), 1135-9.

34. Tatini S.R. Heat-stable nuclease for assessment of staphylococcal growth and likely presence of enterotoxins in foods. *Journal of Food Science*, 1975, 40 (2), 352-6.

35. Billon J., Guerin J., Sebald M. Etude de la toxinogenese de *Clostridium botulinum* type B au cours de la maturation de ages a pate molle. *Le Lait*, 1980, 60, 329-42.

36. Briozzo J., de Lagarde E.A., Chirife J., Parada J.L. *Clostridium botulinum* type A Growth and Toxin Production in Media and Process Cheese Spread. *Applied and Environmental Microbiology*, 1983, 45 (3), 1150-2.

37. Aureli P., Franciosa G., Pourshaban M. Foodborne botulism in Italy. *Lancet*, 1996, 348 (9041), 1594.

38. Simini B. Outbreak of foodborne botulism continues in Italy. *Lancet*, 1996, 348 (9030), 813.

39. International Dairy Federation. Viruses pathogenic to man in milk and cheese, in *Behaviour of Pathogens in Milk*. Ed. International Dairy Federation. Brussels, IDF. 1980, 17-20 IDF Bulletin No. 122

40. Northolt M.D. Growth and inactivation of pathogenic microorganisms during manufacture and storage of fermented dairy products. A review. *Netherlands Milk and Dairy Journal*, 1984, 38 (3), 135-50.

6. FERMENTED MILKS

6.1   Definitions

Fermented milks have been produced by traditional methods for many centuries, and there are several hundred such products recorded around the world. They are produced as a result of microbial 'souring' of milk, usually cows' milk, but also the milk of other species, including sheep, goat, horse and buffalo. Most are very similar, both in terms of their characteristics, and in the technology used to produce them. Many fermented milk products are distinguished only by their region of origin, and very few have become commercially important. Interest in these products, particularly yoghurt, has grown rapidly since the development of flavoured and fruit yoghurts in Europe in the late 1950s, and more recently as a result of the growing demand for, and marketing of, fermented milks as health-promoting foods.

Fermented milks can be conveniently classified on the basis of the type of fermentation they undergo, as lactic, yeast-lactic (e.g. Kefir, Koumiss, acidophilus-yeast milk) and mould-lactic (e.g. Villi). Lactic fermentation products can be further classified, depending on the characteristics of the lactic microflora, as mesophilic (e.g. Filmjolk, Nordic ropy milk, Maziwa lala, Ymer), thermophilic (e.g. Yoghurt, Labneh, Shirkhand, Skyr, Bulgarian buttermilk) and probiotic or therapeutic (e.g. 'Bio'-fermented milks, acidophilus milk, AB-yoghurt, Yakult, Danone, Cultura-AB). Examples of products of each type are given below.

6.2   Lactic Fermentations

6.2.1   Mesophilic

The genera of microorganisms that fall into this category include *Lactococcus*, *Leuconostoc* and *Pediococcus*. The optimal growth temperature is between 25 - 30 °C.

6.2.1.1   Traditional or natural buttermilk

Traditional or natural buttermilk is made from the liquid produced during butter production using a starter culture mixture of *Lactococcus* spp. (*Lactococcus lactis* subsp. *lactis*, and *Lactococcus lactis* subsp. *cremoris*, which are the main producers of lactic acid, and *Lactococcus lactis* biovar *diacetylactis*, the diacetyl flavour-producing organism) and *Leuconostoc mesenteroides* subsp. *cremoris* (also responsible for flavour production).

6.2.1.2 Cultured buttermilk

Cultured buttermilk is also produced mostly using a mixed culture of *L. lactis* subsp. *lactis*, *L. lactis* subsp. *cremoris* and the flavour-producing organisms *L. lactis* biovar *diacetylactis* and *L. mesenteroides* subsp. *cremoris*. It is traditionally made from skimmed milk. Ymer is similar to cultured buttermilk, but differs in the sequence of the manufacturing stages.

6.2.1.3 Nordic sour milks

Nordic sour milks such as Filmjolk and Nordic ropy milk are made using slime-producing *Bacterium lacticus longi*, a synonym of *Lactococcus* spp. (the name *Lactococcus lactis* biovar *longi* has been proposed). The slimy or ropy consistency of the products is also attributed to Butterwort leaves, which are rubbed on the interior of the pails.

6.2.1.4 Cultured cream

Cultured cream or sour cream is made using the same starter cultures as cultured buttermilk, but has a much higher fat content (10 - 40%).

6.2.1.5 Miscellaneous products

Miscellaneous products include a range of traditional products that depend on spontaneous fermentation by naturally present lactic acid bacteria (LAB) in milk. Maziwa lala (commercially called Mala) is made using the same starter culture mixture as buttermilk, but is then sweetened. Susa, made from camel's milk is fermented using hetero-fermentative mesophilic starter cultures. Lben is similar to buttermilk but its production involves spontaneous fermentation. The microflora of this product mainly consists of *L. lactis* biovar *diacetylactis*, *Leuconostoc lactis*, *L. mesenteroides* subsp. *cremoris* and *Leuconostoc mesenteroides* subsp. *dextranicum*; lactobacilli, yeast, mould and coliforms are also present.

6.2.2 Thermophilic

This category encompasses those starter cultures whose growth optimum is between 37 and 45 °C. The genera of microorganisms that fall into this category include *Streptococcus* and *Lactobacillus*.

6.2.2.1 Yoghurt

Yoghurt is a term used to describe a wide range of related products, which may be classified according to legal standards (full-, medium- or low-fat), gel type (set or stirred) and whether or not they are flavoured (natural, fruit, or flavoured) or if they are subjected to a further process (heating, freezing, drying or concentrating). The usual starter culture employed to produce yoghurt is a mixture of *Streptococcus thermophilus* and *Lactobacillus delbrueckii* subsp. *bulgaricus*.

6.2.2.2 Acid buttermilk

Acid buttermilk, also known as Bulgarian buttermilk is made using *L. delbrueckii* subsp. *bulgaricus* as the starter culture. *Str. thermophilus* or a cream culture may also be included in the starter culture.

6.2.3 Probiotic or therapeutic

LAB such as enterococci, lactococci, propionibacteria, *Leuconostoc*, and pediococci are used as probiotics, but the principal organisms are of the bacterial genera *Lactobacillus* and *Bifidobacterium*.

6.2.3.1 Yakult

Yakult is a term for a group of therapeutic products originating from Japan. The starter culture used is *Lactobacillus casei* subsp. *casei* (*L. casei* Shirota), an organism naturally present in the normal intestinal microflora of humans. The organism is a probiotic strain that is thought to have a beneficial effect on the host, by improving the intestinal microbial balance. The positive health benefits of probiotics are reported to be of particular value in the treatment of diseases that result in a disturbance of the intestinal microflora.

6.2.3.2 Acidophilus milk

Acidophilus milk is a traditional therapeutic milk product popular in eastern Europe, but now attracting more attention elsewhere for its perceived beneficial properties. It may be made from skimmed or whole milk, and the starter organism is *Lactobacillus acidophilus*.

6.2.3.3 'Bio' yoghurts

'Bio' yoghurts are made by very much the same process as traditional yoghurt, and are very similar products, but usually use a mixed starter culture consisting of probiotic strains. *Bifidobacterium* spp. are often used, especially *Bifidobacterium bifidum,* and *Bifidobacterium longum,* together with lactobacilli, such as *L. casei* and *L. acidophilus.* These organisms are all found in the normal intestinal microflora and are considered to have a beneficial effect on human health.

6.3   Yeast - Lactic Fermentations

Mesophilic LAB, thermophilic LAB and yeast are the main fermentation genera.

6.3.1   Kefir

Kefir is a rather foamy and effervescent fermented milk that contains about 1% lactic acid and 0.5 - 1.0% alcohol, and is popular in eastern Europe and Mongolia. The starter culture consists of small, white 'kefir grains', about 2 - 10 mm in diameter. These grains contain a complex and quite variable microbial community, but little is known about how they develop. The grains usually contain LAB such as *Lactobacillus* spp. (*L. delbreuckii* subsp. *bulgaricus*) plus *Lactococcus* spp. (*L. lactis* subsp. *lactis*), *Leuconostoc* spp., and *Str. thermophilus,* acetic acid bacteria (*Acetobacter aceti* and *Acetobacter rasens*), contaminants such as mould (*Geotrichium* spp.), and a number of yeast species such as *Saccharomyces and Kluyveromyces,* but the principal yeast species present is *Candida kefir* (synonym: *Candida kefyr*; telemorph: *Kluyveromyces marxianus*).

6.3.2   Koumiss

Koumiss is traditionally made in central Asia from mares' milk, but is now often made from skimmed, or whole cows' milk with added sugar. Starter cultures contain LAB such as lactobacilli (*L. delbrueckii* subsp. *bulgaricus* and *L. acidophilus*), strains of lactose-fermenting yeasts (*K. marxianus, Saccharomyces* spp., *Torula koumiss*), non-lactose-fermenting and non-carbohydrate-fermenting yeasts. The finished product contains lactic acid, alcohol and carbon dioxide, producing a slightly effervescent drink.

6.3.3   Miscellaneous Products

Miscellaneous products such as acidophilus-yeast milk fall under the yeast-lactic group of fermented products, but little is known about the technology of these beverages.

6.4   Mould - Lactic Fermentations

Mesophilic LAB and mould are the genera responsible for fermentation.

6.4.1   Villi

Villi is a fermented milk product from Finland, which is made from whole milk, using a starter culture of *L. lactis* subsp. lactis biovar *diacetylactis*, *L. mesenteroides* subsp. *cremoris*, and the mould *Geotrichum candidum*. The mould grows on the layer of fat that forms on the top of the product and produces a felt of mycelium.

6.5   Initial Microflora

The initial microflora of fermented milk products is determined largely by the microflora of the whole and skimmed milks from which they are made.

6.6   Processing and its Effects on the Microflora

Although there is a very wide range of fermented milk products, the manufacturing technology used is generally very similar. The principal differences are in the starter cultures used, the composition and treatment of the milk, and the fermentation conditions. Therefore, for the purposes of this chapter, yoghurt manufacture is used as a representative example of fermented milk processes, since yoghurt is the most commercially important of these products. An outline of the process is depicted in Figure 6.1.

6.6.1   Initial processing

Several different varieties of yoghurt are produced, but the three main types are: set; stirred; and drinking yoghurt. Yoghurt is most commonly made from cows' milk, but may also be produced from the milk of sheep, goats, and, occasionally, other animal species.

The composition of yoghurt varies slightly, and in some countries is regulated by legislation. Both whole milk and reduced-fat milks are used to produce yoghurt, but reduced-fat products have the largest market share in most countries. A fat content of approximately 1.5% is typical for a low-fat yoghurt, and the milk is usually standardised to control the final fat content. The protein composition and quality of the milk are important, since they may have a significant effect on texture. Only milk of good microbiological quality should be used, in order to avoid problems of proteolysis associated with bacterial activity, and the production of bacterial proteases by psychrotrophs. These enzymes may significantly alter the physical properties of the yoghurt, and cause defects.

The milk-solids-not-fat (MSNF) content of the milk is usually increased to give a higher viscosity in the finished product. This may be done by fortification with non-fat dried milk or other dairy powders, or by concentration methods, such as evaporation under vacuum, or by membrane filtration (ultra filtration or reverse osmosis). For most yoghurt, an MSNF level of about 15% is typical, but for drinking yoghurt, levels of less than 11% are preferred. The milk is usually then filtered, to remove undissolved particles, de-aerated to provide conditions that favour rapid starter growth, and homogenised to improve texture and help prevent syneresis. Stabilisers may also be added to stirred yoghurts to improve viscosity and further reduce the likelihood of syneresis. Pre-gelatinised starch or plant gums are the most commonly used stabilisers.

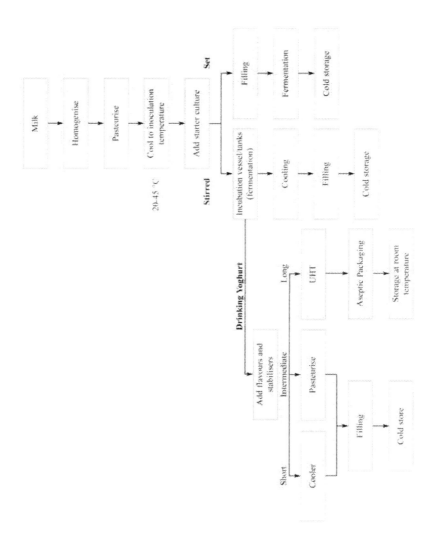

Fig.6.1. Production of stirred and set fermented milk

6.6.2 Heat treatment

A heat treatment is generally applied to milk for yoghurt manufacture. A process of 80 - 85 °C for 30 minutes is typical for batch processes, but, for continuous processes, a heat treatment of 90 - 95 °C for 5 - 10 minutes is more usual. In some cases, a full ultra high temperature (UHT) process (133 °C for 1 second) may be applied. This relatively severe heat treatment has a number of effects. Vegetative bacterial cells, which may include pathogens such as *Salmonella*, are killed, leaving only heat-resistant bacterial spores. Non-pathogenic organisms that might interfere with the growth of the starter culture are therefore reduced to very low levels. The heat treatment also has a significant effect on the final viscosity of the yoghurt, and improves texture by causing denaturation of the whey proteins (albumins and globulins) and the three-dimensional aggregation of casein molecules. The oxygen concentration of the milk is also further reduced, improving conditions for starter culture growth, since the starter organisms are generally microaerophilic. Finally, starter culture activity can either be stimulated or inhibited due to the breakdown products of heat-damaged milk proteins.

6.6.3 Fermentation

After heat treatment, the milk is cooled to 30 or 45 °C and is then inoculated with the starter culture (using long or short incubation periods, respectively). Most commercial yoghurt production now uses a mixed inoculum containing defined strains of *Str. thermophilus* and *L. delbrueckii* subsp. *bulgaricus*. During fermentation, the two starter organisms grow synergistically. Initially, *Str. thermophilus* grows rapidly and produces lactic acid. As the acidity rises, so *L. delbreuckii* subsp. *bulgaricus* becomes more active and produces more acid along with aroma compounds. It has been found that the growth of *Str. thermophilus* stimulates the growth of the *Lactobacillus*, probably by the production of a growth factor, thought to be formic acid, which stimulates proteolytic activity. *Lactobacillus* growth is further stimulated by the production of low levels of carbon dioxide (CO_2) from urea (30 - 50 mg CO_2/kg of milk within the first hour of incubation). The lactobacilli, in turn, stimulate the growth of *Str. thermophilus* by releasing peptides and, to lesser extent, amino acids as by-products of proteolysis.

The specific strains used are chosen for their effect on product flavour and texture, and strains may be used in rotation to prevent occasional problems with bacteriophage infection. Good hygienic practices after pasteurisation are required to prevent build-up of bacteriophage pools in stagnant whey. It is important that the inoculum contains a balanced population of the two organisms, usually a 1:1 ratio. Both liquid and freeze-dried cultures are used. Most commercial manufacturers use an inoculum of about 2 - 3% v/v. Incubation of the inoculated milk at 40 - 45 °C produces a rapid fermentation, which is normally complete within 3 - 4 hours. During this time, lactic acid is produced, giving an eventual acidity of 0.9 - 0.95% and a pH of approximately 4.6 - 5.0. The starter organisms may also produce aroma compounds, such as acetaldehyde, acetone, acetoin and diacetyl, and exopolysaccharides, which improve texture and viscosity, although if too much extracellular material is produced, a 'ropy' texture fault may result. Set yoghurt is fermented in the final container, but stirred yoghurt is fermented in bulk, and stirred slowly, for only a few minutes, during the process.

6.6.4   Cooling and packing

When fermentation is complete, the yoghurt is initially cooled to about 15 - 20 °C, to minimise further acid production. At this point, sweeteners, flavours and/or fruit purées may be added. These additions must be of good microbiological quality, since, for most yoghurts, no further processing is applied (for set yoghurts, additional ingredients must be added before fermentation). The product is then dispensed into the final containers (stirred yoghurts) and further cooled to < 5 °C. A shelf life of about 3 weeks at this temperature is typical, although acid continues to be produced during this time and may affect flavour. Some yoghurt is heat treated after fermentation to destroy starter organisms, and this increases shelf life to several months. Greek-style, concentrated yoghurts are produced by further separation of the yoghurt after fermentation to increase the fat and solids content. Yoghurt may also be frozen, or dried for use as an ingredient.

6.7   Probiotic products

The production of fermented milks with probiotic organisms is very similar to the basic process for yoghurt, and uses the same key stages. However, the characteristics of the organisms used require that certain modifications be made. Since probiotic bacteria are intended to colonise the gastrointestinal tract of consumers, it is important that the number of viable cells in the product be as high as possible. Direct vat inoculation of probiotic starters is common, using cultures in nutrient-supplemented milk. Bulk starter media may also be used, especially for *Bifidobacterium* spp., which can be difficult to grow. Inoculation rates also tend to be higher for probiotic starters. Rates of 10 - 20% v/v can be used. *Bifidobacteria* are usually used in combination with other LAB as they may produce quantities of acetic acid, which can adversely affect flavour, and to overcome slow acid production.

   L. acidophilus is a slow-growing organism and therefore requires a long incubation time to produce sufficient acid. This means that *L. acidophilus* fermentations are likely to be disrupted by spore-forming bacteria in the early stages if these organisms are present in significant numbers. For this reason, the milk is usually given a severe heat process such as 95 °C for one hour, or a UHT treatment, to reduce spore levels.

   A further challenge is ensuring the survival of sufficient numbers of the probiotic organisms in the product throughout its shelf life for the product to be classified as having probiotic properties. Survival is influenced by various factors, including acidity, pH, temperature, and oxygen concentration. Therefore, careful control of these factors during processing and storage is important, as is the initial probiotic strain selection. Recent studies suggest that it is possible to maintain high numbers of viable cells throughout shelf life (1, 2).

6.8   Spoilage

6.8.1   Bacteria

Bacterial spoilage of fermented milks is unlikely. The combination of severe heat treatment, the presence of active starter organisms, low pH and chilled storage prevents the growth of bacterial contaminants. However, if acid production is slow, sporeforming organisms and post-process contaminants may be able to grow. If products are stored at too high a temperature, continued acid production and proteolysis by starter organisms may impair flavour and cause bitterness. If bacteriophage contamination occurs post heating, they may inactivate the starter completely or cause slow acidification. Occasionally, rapid growth of non-starter LAB, present as post-heat treatment contaminants, may also lead to over-acidification and undesirable flavours. It is therefore important to manufacture fermented milks in hygienic conditions to minimise recontamination.

Spoilage of fermented milks, in general, is likely to be caused by yeasts, or by moulds, and most documented cases of spoilage refer to yoghurts.

6.8.2   Yeasts

Acid-tolerant, psychrotrophic, fermentative yeasts may be able to grow in yoghurt, and cause blowing as a result of the production of CO_2. Yoghurts that contain sucrose through the addition of fruit purées, flavours, or chocolate, are particularly vulnerable to this kind of spoilage, since many yeasts are able to ferment sucrose. Species of *Candida*, *Saccharomyces*, *Pichia*, *Rhodotorula*, *Kluyveromyces*, and *Torulopsis* have been isolated from blown packs of yoghurt (3). Fruit purées, cereals, honey, nuts and spices are the most likely source of yeast contamination. Purées used in yoghurt should not contain any detectable yeasts at a level of 1 g, since even low numbers of cells may reduce shelf life. Other sources of contamination are airborne dust particles, contaminated packaging material, poor hygiene on processing lines and other ingredients.

Few yeasts are able to ferment lactose, and therefore natural yoghurt without additions is much less prone to yeast spoilage. The exceptions are *Kluyveromyces marxianus* var. *lactis* and *Kluyveromyces marxianus* var *marxianus*, which will ferment lactose, and are the most common cause of natural yoghurt spoilage. This organism may colonise equipment and surfaces in the yoghurt manufacturing plant, and effective cleaning and hygiene procedures are needed to prevent contamination (4). This species may also grow at high temperatures, and, if present as a contaminant in a starter culture, could disrupt fermentation.

6.8.3   Moulds

Growth of acid-tolerant moulds in yoghurt is restricted by lack of oxygen, and agitation of the product during transport tends to suppress growth at the surface. However, growth of species of *Mucor, Aspergillus, Rhizopus, Alternaria* and *Penicillium* can occur at the product/air interface during retail storage, leading to the production of mycelial mats or buttons, and visible spoilage. Mould spores may contaminate yoghurt via airborne dust particles, packaging and added ingredients, as well as through poor hygiene (5).

6.9   Pathogens: Growth and Survival

Fermented milks have a good safety record in terms of foodborne disease, and there are very few recorded incidents of food poisoning associated with these products. Milk used in fermentations is generally subjected to a severe heat treatment sufficient to destroy vegetative pathogens. Furthermore, it is generally considered that pathogens are not able to grow in fermented milks, and that their survival is likely to be limited by the low pH, high acid concentrations and presence of inhibitory compounds such as bacteriocins. For example, *Campylobacter* spp. are rapidly killed in the presence of lactic acid.

Experiments to determine the survival of foodborne pathogens in yoghurt and other fermented milks tend to produce quite variable results. Survival times can be influenced by pH, acidity and the characteristics of the starter culture used. Salmonellae are usually considered to be inactivated at levels of lactic acid of > 1% (6), but it is possible that pathogens may adapt to acid conditions over time, and the effect of this adaptation on survival should be considered. There is also a trend towards the development of milder-flavoured products with significantly lower levels of acid. Pathogen survival in these products may be significantly enhanced. Indeed, the length of time over which viable *Salmonella typhimurium* cells could be recovered from inoculated fermented milk was found to increase at lower levels of acid production (7). The demand for fermented milk has lead to the manufacture of these products in unapproved premises; as was highlighted in March 2007 by a Food Alert issued by the Food Standards Agency. Therefore, it is not advisable to rely on low pH and acid production to ensure product safety; effective hygiene procedures to prevent pathogen contamination during processing are also necessary.

6.9.1   Listeria monocytogenes

It is generally considered that *L. monocytogenes* is unlikely to be able to grow in fermented milks, but survival in the finished product is possible.

The behaviour of *L. monocytogenes* in fermented milks has recently been reviewed (8). It has been found that the organism may be able to grow in some buffered culture media used for the preparation of starters, and that contaminated starter cultures are a potential source of *Listeria* in the finished product. Studies with both cultured buttermilks and yoghurts inoculated with *L. monocytogenes* before fermentation showed that survival was influenced by starter culture type, fermentation temperature and final acidity. In some fermented buttermilks, viable cells could be recovered after twelve and a half weeks of refrigerated storage (9). Survival times in yoghurt have been found to be shorter, and, in general, the lower the pH of the finished product, the shorter the survival time. Survival of *L. monocytogenes* inoculated into yoghurt after fermentation (possibly a more realistic scenario) has also been investigated. Survival for up to 3 weeks was recorded, although the majority of the cells were inactivated in the first 12 days (10). A UK survey of 100 samples of retail and farm-produced yoghurts showed that all the samples were negative for *L. monocytogenes* (11).

6.9.2   *Escherichia coli*

In general, *E.coli* is rapidly inactivated by lactic fermentation; a study showed that rapid inactivation of *E.coli* occurred in 4 days at 7.2 °C when it was added to yoghurt samples (12). However, the unusual acid tolerance of verotoxigenic *E.coli* O157:H7 is of concern, and in 1991 an outbreak occurred in north-west England, associated with locally produced live yoghurt. The organism could not be isolated from the yoghurt or milk, but epidemiological evidence indicated a link (13). Recent studies have demonstrated that *E.coli* O157:H7 inoculated into commercial yoghurt and other fermented milks, survived for up to 12 days in yoghurt, and for several weeks in sour cream and cultured buttermilk and that the addition of sugar to cultured milk products enhances survival of *E.coli* O157:H7 (14). Studies have also shown that *E.coli* O157:H7 capable of producing colonic acid persist longer in yoghurt (15). Contamination of these products with the organism is therefore a potential health hazard, since the infective dose is thought to be low (16).

6.9.3   *Staphylococcus aureus*

Staph. aureus is very unlikely to grow in fermented milks; however, a case of staphylococcal food poisoning was reported in 1970. The cause was attributed to the high sugar content of the product, which favoured *Staph. aureus* growth and toxin formation, while inhibiting the starter culture (lactic acid) (12). Survival in inoculated sour cream, cultured buttermilk and yoghurt has also been shown. In sour cream inoculated at a level of 10^5 cells/g, viable cells could be recovered after 7 days, but this was not the case at lower inoculation rates (17). The survival of *Staph. aureus* during fermentation and subsequent storage has also been studied, with similar results. At high inoculation rates, viable cells survived through fermentation, but died out during chilled storage (18).

6.9.4   *Clostridium botulinum*

In 1989, there was a well-documented outbreak of botulism in the UK associated with hazelnut yoghurt. The contamination was not as a result of a problem with the manufacture of the yoghurt itself, but with underprocessed hazelnut purée, added as a flavouring. The purée had been prepared with artificial sweeteners instead of sugar. As a result, the raised water activity allowed *C. botulinum* spores to germinate and produce toxin (19). Although an unusual case, this incident emphasises the importance of controlling the microbiological quality of those ingredients added after fermentation. Proper application of HACCP principles to new product development processes should minimise the risk of problems like this occurring.

6.9.5   *Yersinia/Aeromonas spp.*

The ability of these organisms to grow at low temperatures suggests that their presence in fermented milks could be a hazard. The growth and survival of both organisms in yoghurt have been investigated. *Aeromonas hydrophila* was found to be completely inhibited after 5 days of refrigerated storage, but *Yersinia enterocolitica* could still be detected at the end of shelf life after 26 days (20). However, as with other pathogens, survival through fermentation and storage is probably dependent on the rate of acid production and the final pH. A later study determined survival times of only 5 days for *Y. enterocolitica* during chilled storage (21).

6.9.6   *Bacillus cereus*

Spore germination and growth of *B. cereus* in fermented milks are prevented by low pH. However, growth of *B. cereus* has been shown in yoghurt milk at 31 °C, although, as the pH dropped, the growth rate declined, and it ceased at pH 5.7. Although it is possible that high levels could be reached when initial acid production is slow, *B. cereus* is not normally considered a hazard in fermented milks (22).

6.9.7   *Toxins*

If the milk used to produce yoghurt and other fermented milks is contaminated with mycotoxins, probably through contaminated animal feed, it is possible that the finished product will also be contaminated. It has been shown that aflatoxins are stable during the manufacture of yoghurt and subsequent chilled storage for 21 days (23).

Concern has also been expressed regarding mycotoxigenic moulds growing on the surface of yoghurt, following the isolation of the toxigenic species *Penicillium frequentans* as a contaminant in a commercial yoghurt sample (24). However, since mycotoxin production would be expected to coincide with visible growth, and visibly spoiled products are unlikely to be consumed, this does not seem to be a serious hazard.

6.10   Probiotic Products

Since many probiotic cultures used to ferment milk are slow acid producers, it may be that there is an increased opportunity for contaminating pathogens to grow to dangerous levels before the pH drops to inhibitory levels. For this reason, it becomes even more important to implement effective hygiene procedures to ensure that potential pathogens are not able to contaminate ingredients or the processing environment.

Concern has also been expressed over the safety of some probiotic cultures, particularly strains of *Enterococcus faecium*, which may be an opportunistic pathogen, and display multiple antibiotic resistance. Therefore, considerable care must be exercised in the selection of probiotic organisms, to ensure that they do not present any discernible health risk to consumers.

6.11   References

1. Shin H.-S., Lee J.-H., Pestka J.J., Ustunol Z. Viability of bifidobacteria in commercial dairy products during refrigerated storage. *Journal of Food Protection*, 2000, 63 (3), 327-31.

2. Schillinger U. Isolation and identification of lactobacilli from novel-type probiotic and mild yoghurts and their stability during refrigerated storage. *International Journal of Food Microbiology*, 1999, 47 (1-2), 79-87.

3. Kosse D., Seiler H., Amann R., Ludwig W., Scherer S. Identification of yoghurt-spoiling yeasts with 185 rRNA-targeted oligonucleotide probes. *Systematic and Applied Microbiology*, 1997, 20 (3), 468-80.

4. Fleet G.H. Yeasts in dairy products. A review. *Journal of Applied Bacteriology*, 1990, 68 (3), 199-211.

5. Filtenborg O., Frisvad J.C., Thrane U. Moulds in food spoilage. *International Journal of Food Microbiology*, 1996, 33 (1), 85-102.

6. Hobbs B.C. General aspects of food poisoning and food hygiene. *Journal of the Society of Dairy Technology*, 1972, 25 (1), 47-50.

7. Park H.S., Marth E.H. Behaviour of *Salmonella typhimurium* in skim milk during fermentation by lactic acid bacteria. *Journal of Milk and Food Technology*, 1972, 35 (8), 482-8.

8. Ryser LT. Incidence and behavior of *Listeria monocytogenes* in cheese and other fermented dairy products, in *Listeria,Listeriosis and Food Safety*. Eds. Ryser LT., Marth LH. New York, CRC Press. 2007, 405-502.

9. Schaack M.M., Marth E.H. Survival of *Listeria monocytogenes* in refrigerated cultured milks and yogurt. *Journal of Food Protection*, 1988, 51 (11), 848-52.

10. Choi H.K., Schaack M.M., March E.H. Survival of *Listeria monocytogenes* in cultured buttermilk and yoghurt. *Milchwissenschaft*, 1988, 43 (12), 790-2.

11. Kerr K.G., Rotowa N.A., Hawkey P.M. *Listeria* in yoghurt? *Journal of Nutritional Medicine*, 1992, 3 (1), 27-9.

12. International Commission on Microbiological Specifications for Foods. Milk and dairy products. (Microorganisms in milk and dairy products.), in *Microorganisms in Foods 6: Microbial Ecology of Food Commodities*. Ed. International Commission on Microbiological Specifications for Foods. New York, Kluwer Academic/Plenum Publishers. 2005, 643-715.

13. Morgan D., Newman C.P., Hutchinson D.N., Walker A.M., Rowe B., Majid F. Verotoxin producing *Escherichia coli* O157 infections associated with the consumption of yoghurt. *Epidemiology and Infection*, 1993, 111 (2), 181-7.

14. Chang J.H., Chou C.C., Li C.E. Growth and survival of *Escherichia coli* O157:H7 during the fermentation and storage of diluted cultured milk drink. *Food Microbiology*, 2000, 17 (6), 579-87.

15. Lee S.M., Chen J. Survival of *Escherichia coli* O157:H7 in set yoghurt as influenced by the production of an exopolysaccharide, colanic acid. *Journal of Food Protection*, 2004, 67 (2), 252-5.

16. Dineen S.S., Takeuchi K., Soudah J.E., Boor K.J. Persistence of *Escherichia coli* O157:H7 in dairy fermentation systems. *Journal of Food Protection*, 1998, 61 (12), 1602-8.

17. Minor T.E., Marth E.H. Fate of *Staphylococcus aureus* in cultured buttermilk, sour cream, and yoghurt during storage. *Journal of Milk and Food Technology*, 1972, 35 (5), 302-6.

18. Pazakova J., Turek P., Laciakova A. The survival of *Staphylococcus aureus* during the fermentation and storage of yoghurt. *Journal of Applied Microbiology*, 1997, 82 (5), 659-62.

19. O'Mahony M., Mitchell E., Gilbert R.J., Hutchinson D.M., Begg N.T., Rodhouse J.C, Morris J.E. An outbreak of food borne botulism associated with contaminated hazelnut yoghurt. *Epidemiology and Infection*, 1990, 104 (3), 389-95.

20. Aytac SA, Ozbas Z.Y. Survey of the growth and survival of *Yersinia enterocolitica* and Aeromonas hydrophila in yogurt. *Milchwissenschaft*, 1994, 49 (6), 322-5.

21. Bodnaruk P.W., Williams R.C, Golden D.A. Survival of *Yersinia enterocolitica* during fermentation and storage of yoghurt. *Journal of Food Science*, 1998, 63 (3), 535-7.

22. Robinson R.K., Tamime A.Y., Wszolek M. Microbiology of fermented milks, in *Dairy Microbiology Handbook: The Microbiology of Milk and Milk Products*. Ed. Robinson R.K. New York, John Wiley and Sons. 2002, 367-430.

23. Blanco J.L., Carrion B.A., Liria N., Diaz S., Garcia M.E., Dominguez L., Suarez G. Behaviour of aflatoxins during manufacture and storage of yoghurt. *Milchwissenschaft*, 1993, 48 (7), 385-7.

24. Garcia A.M., Fernandez G.S. Contaminating mycoflora in yoghurt: General aspects and special reference to the genus Penicillium. *Journal of Food Protection*, 1984, 47 (8), 629-36.

7. ICE CREAM AND RELATED PRODUCTS

7.1   Definitions

Cream ices are frozen dairy desserts containing milk fats. Their composition is regulated by legislation in many countries, and varies considerably. In the United States, ice creams must typically contain fat levels of 10% or more (12% for "premium ice cream" and 14% for "super premium ice cream"). In the UK, ice creams must contain no less than 5% fat and 2.5% milk solids. Additional flavourings and ingredients such as nuts and chocolate are often added to create a range of ice cream varieties. Examples of such products include crème glacée, eiskrem, and crema di gelato.

Ice cream must also meet minimum fat requirements, but may contain milk fat, vegetable fats, or non-dairy animal fats, Such products include mellorine (used in the US) Ijs (from the Netherlands) and glaces de consummation (Belgium). Countries like France and Germany, prohibit the use of non-dairy fat in ice cream. In the UK, non-dairy fat is permitted in ice cream, but 'dairy ice cream' is used to describe those products made exclusively from milk fat.

Milk ices are made using milk, but without additional fat. They contain less fat than ice cream (3 - 5%), but higher levels of sugar and non-fat milk solids, e.g. glace au lait, milcheis, gelato al latte.

Custards or French ice creams or French custard ice creams are similar to milk ices, but also contain at least 1.5% added egg yolk solids.

Ices or water ices are made with fruit juices and/or pulp and water. They may also contain sugar, acid (for example, citric, malic or tartaric), stabilisers (e.g. gelatin, pectin), colour and flavour. These products may be frozen with or without agitation and the incorporation of air. 'Ice lollies' are water ices frozen without agitation. Examples of agitated products include 'Frappe' made in 'slush' conditions, and 'punch' made with alcoholic liquid instead of water.

Sherbet is similar to water ice, but also contains small quantities of ice cream, liquid milk, milk fat and milk solids. Air is often incorporated into the product during freezing.

Sorbets are also similar to water ices but have a high content of sugar, fruit and fruit juice. In addition, the product contains stabilisers and egg white, and has an overrun of 20% or less. Sorbets often contain exotic flavours.

Mousse is a flavoured, frozen whipped cream, to which stabilisers are added to maintain texture.

Cassatas are made in round moulds and have various flavoured layers of ice cream. They may also have fruit, liqueurs, chocolate, nuts or slices of sponge cake (sometimes soaked in liqueur).

Splits are made on a stick; the core consisting of ice cream and the outer layer made of fruit water ice or chocolate with nuts and/ or biscuit crumbs.

Frozen yoghurts are made by freezing a pasteurised mix of milk fat, milk solids non-fat (MSNF), sweeteners, stabilisers, and yoghurt (10 - 20%). They may be flavoured with fruit puree.

Other types of ice creams beginning to enter the market or triggering research interest are ice creams with different fat contents, probiotic ice creams and novelties;

Ice creams with different fat contents include reduced fat (25% less fat), low-fat (with no more than 3g of fat), and non-fat (less than 0.5g fat)

Probiotic ice creams have been investigated, and studies have shown that ice creams could be used to deliver probiotic bacteria, without any change in sensory properties.

Novelty products are generally defined as 'unique single-serve portion-controlled products' made from ice cream with special flavours and confectionery. They may be shaped and enrobed in chocolate or water ice, and/or moulded onto a stick or available as cup items e.g. coated ice cream bars, ice cream cakes and logs.

7.2   Initial Microflora

The initial microflora of ice cream prior to pasteurisation is largely determined by the individual ingredients, milk, cream, dried milk, etc. Where flavourings and other ingredients, such as sugar, nuts, fruit and chocolate, are added, this is usually done after pasteurisation. Therefore, there is the potential for such additions to introduce a wide range of other organisms not usually found in dairy products. This must be carefully considered, as it is a potential source of pathogenic organisms.

7.3   Processing and its Effects on the Microflora

A schematic outline of ice cream production is shown in Figure 7.1.

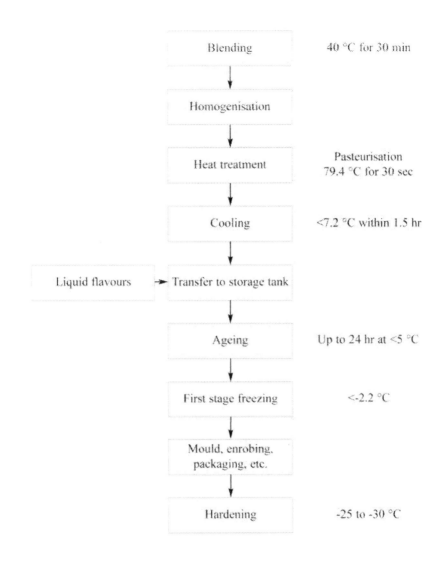

Fig. 7.1. Production of ice cream

7.3.1 *Ingredients*

7.3.1.1 *Fresh whole milk*

Fresh whole milk is a good source of fat and non-fat milk solids (NFMS) for the manufacture of milk ices, but, for ice cream, both fat and NFMS levels must be increased by supplementation with other ingredients. It is important that fresh milk used for ice cream has been properly pasteurised, stored correctly to minimise the growth of psychrotrophs, and used quickly.

7.3.1.2 Fresh cream

Fresh cream is the best source of additional milk fat, but it is both costly and highly perishable. Alternatives include unsalted butter, sweet cream, and anhydrous milk fat (butter oil). Where non-dairy fats are permitted, partly hydrogenated vegetable oils are often used, particularly palm oil, palm kernel oil and coconut oil. Highly processed fats and oils are unlikely to carry significant microbial contamination, but butter may contain lipolytic bacteria such as *Pseudomonas fragi*, which may cause tainting.

7.3.1.3 Additional NFMS

Additional NFMS (which include sugars, proteins and minerals) can be obtained using concentrated liquid skimmed milk, sweetened condensed milk, dried skimmed milk powder, whey powders and modified liquid whey concentrates. Sweetened condensed milk and whey powders may lead to the formation of large lactose crystals that may result in a 'sandy' texture defect. Spray-dried whole milk powders are sometimes used to add both NFMS and milk fat to ice cream, but these products are vulnerable to the development of off-flavours and rancidity. Skimmed milk powders may sometimes be contaminated by large numbers of *Bacillus* spores, including *Bacillus cereus*. This is undesirable, both from a public health point of view, and because psychrotrophic bacilli may be able to grow in the ice cream mix and cause eventual spoilage. Dried milk may serve as a source of *Listeria monocytogenes*, as these organisms are known to survive the spray-drying process.

7.3.1.4 Sugars

Sugars are used to sweeten most ice creams, and this also increases the total solids content of the mix. Sucrose is most commonly added, but glucose syrups and dextrose powder are also used, sometimes in combination with sucrose. Few microbiological problems are anticipated with these ingredients, although syrups may support the growth of some osmophilic yeasts (*Zygosaccharomyces*, *Candida*, *Pichia*, *Torula*), and surface mould growth is also possible. Nowadays, fructose or artificial sweeteners are being used to manufacture diabetic ice cream; the safety and quality of the product may be compromised as the bacterial growth inhibitory effects of artificial sweeteners may not be as effective as those exerted by sugar. Bacteria present could grow before freezing.

7.3.1.5 Stabilisers

Stabilisers are added to most ice cream mixes to increase viscosity and give the product the correct texture. A number of different stabilisers can be used, and the most commonly added to ice cream are alginates, carrageenan, carboxymethyl cellulose, and gums (locust bean, guar and xanthan). Emulsifiers are also added to give the ice cream a smooth texture by preventing the agglomeration of fat globules, and helping to produce smaller air cells during processing. Egg yolk was traditionally used for this purpose, only eggs that have been pasteurised should be used if eggs are added after heat treatment. Glyceryl monostearate, polyoxethylene glycol, and sorbitol esters are now more common, although not all of these are permitted in some countries. These materials should not present any significant microbiological problems, but should be obtained from a reputable source.

7.3.1.6 Colours and flavours

Colours and flavours, such as vanilla and chocolate, are also incorporated into most ice cream formulations. Synthetic colours and flavours are now being replaced by natural or 'nature-identical' versions in response to consumer preference. Other value-added ingredients, such as nuts, chocolate chips, and fruit pieces, may also be added. Most flavours are added after pasteurisation, and their microbiological quality is therefore very important, as is the standard of hygiene used in the storage and handling of these ingredients. For example, fruits may support high levels of yeast populations, and nuts may be contaminated by xerophilic moulds, some of which could be mycotoxin producers. Some natural flavouring ingredients, such as coconut and raw spices, are possible sources of pathogens, including *Salmonella*, and should be heat-treated if possible.

Air incorporated into the product must be processed (i.e. filtered) to ensure that it is not contaminated.

7.3.2 Mixing

The calculation of the mix formulation is dependent upon the type of product being manufactured, but it is also influenced by the type of freezing equipment used, and the need to obtain a finished product that has the correct fat to sugar, and solids to water ratios, to give an acceptable texture. Small manufacturers may mix each batch manually in the pasteurisation tank, but in larger operations, the addition of ingredients to each batch by weight or volume may be automated, and a number of batch blending tanks may be used to ensure a continuous flow of mix to the pasteuriser. The hydrated mix is likely to provide suitable conditions for rapid microbial growth, especially if some pre-heating is necessary to disperse dry ingredients. It may be necessary to hold the batch briefly to allow the stabiliser to hydrate, but pasteurisation should generally be carried out as quickly as possible. Excessive microbial growth before pasteurisation could cause tainting, and, in extreme cases, might compromise the effectiveness of the thermal process.

7.3.3   Heat treatment

The heat treatment of ice cream mixes is often defined in national legislation, and varies slightly from country to country. The stipulated processes are usually based on those applied to milk, but are generally higher. This is to allow for the protective effect of the mix on microbial cells, which confers a higher heat resistance than would be the case in fresh milk. For example, it has been shown that the heat resistance of *L. monocytogenes* is increased by some of the ingredients used in ice cream mixes, particularly stabilisers. D-values at 54.4 °C for *L. monocytogenes* in ice cream mix were approximately four to six times those obtained in milk (1). The minimum recommended pasteurisation requirement for ice cream mixes in the UK are not less than 65.6 °C for at least 30 minutes, 71.1 °C for at least 10 minutes, or 79.4 °C for at least 15 seconds (2). Ice cream pasteurisation destroys most vegetative cells and is sufficiently severe to reduce microbial counts to 500/g or less. Most of the survivors are bacterial spores. A sterilised ice cream mix can be obtained by heating the mix to no less that 148 °C for at least 2 seconds (2).

Small processors may use low-temperature, long-time pasteurisation (LTLT) conditions in a batch process, but most manufacturers now use high-temperature, short-time (HTST) conditions in plate heat exchangers. Ultra high temperature (UHT) processing may also be applied by direct steam injection, or in scraped surface heat exchangers. One problem with these continuous processes is the very viscous nature of ice cream mixes, which may cause fouling of surfaces in heat exchangers, but may also affect the flow characteristics of the mix. If conditions of laminar, rather than turbulent flow are established, there is a possibility of underprocessing. This effect has been demonstrated for ice cream mixes during HTST processing (2). As the pasteurisation of the ice cream mix is essential for product safety and microbiological quality, it is extremely important to ensure that the mix receives an adequate heat treatment.

7.3.4   Homogenisation

The size of the fat globules in the mix must be reduced during processing to improve the whipping and air incorporation properties of the product. This is usually done by homogenisation. The homogeniser is often incorporated into the pasteurising equipment and may act as the metering unit for the HTST pasteuriser. In some cases, homogenisation is carried out downstream of the pasteuriser. This may cause microbiological problems as a result of the complexity of homogenisers, which are difficult to clean and sanitise effectively, and may act as sites of recontamination for heat-treated mix. It is recommended that homogenisation be carried out before, or during, pasteurisation wherever possible.

7.3.5 Cooling and ageing

In the UK, after pasteurisation it is recommended that the mix is cooled as quickly as possible to no more than 7.2 °C within a maximum time of 1.5 hours. This recommendation does not apply if the mix is sterilised and then transferred immediately to a sterile airtight container under aseptic conditions, and the container remains unopened prior to freezing (18). The mix is then held at that temperature for a time before freezing. This process is known as ageing, and is necessary to allow further physical changes to occur. During ageing, stabilisers and milk proteins hydrate further, and fat crystallisation occurs. Ageing should normally be completed within 24 hours, since longer holding times present a risk of psychrotrophic growth, either by spore-formers that have survived pasteurisation or by post-process contaminants. This may result in spoilage of the mix before freezing. Adequate temperature control during ageing is critical, as is effective cleaning of storage tanks and processing equipment to minimise recontamination of the mix.

7.3.6 Freezing

Ice cream freezing is usually a two-stage process. In the first stage, which may be a batch or continuous process, the mix is cooled to at least -2.2 °C (preferably -5 to -10 °C) whilst air is incorporated into it. If temperatures rise above -2.2 °C the product must be reheated. The incorporation of air in the frozen mix causes an increase in volume (known as the overrun). The overrun varies and may be up to 100%, depending on the nature of the product. It has been shown that freezing using batch freezers results in significant destruction of bacterial cells, probably through mechanical damage caused by ice crystals, but, in continuous systems, which freeze more rapidly, the destructive effect is much less marked (4). Effective cleaning and sanitation of ice cream freezers are important to prevent recontamination of the mix during freezing. Many designs are difficult to clean thoroughly, although large-scale continuous freezers may now incorporate clean-in-place (CIP) systems.

After the initial freezing process, the ice cream may be packed directly into the final packaging, shaped in a mould, frozen onto a stick, coated or enrobed in chocolate, or may have other ingredients, such as nuts, added. The product is then immediately cooled further to -25 to -30 °C by the second stage of freezing, referred to as hardening. This is carried out either in freezing tunnels or in hardening rooms. If necessary, further final packaging is then applied and the product is stored at about -25 °C or less. Once the ice cream is frozen hard (core temperature of -18 °C), all microbial growth is prevented. However, the finished product must be of a high microbiological standard, as many pathogens are able to survive for long periods in ice cream. For example, *Salmonella* has been shown to survive for 7 years in ice cream (5).

7.4   Distribution

Although no microbial growth can occur in ice cream during storage, there is an opportunity for further contamination to occur at the point of sale. This is particularly the case with bulk products that are dispensed by an operative and presented to the consumer, such as ice cream sold in cones. Microbial contamination may come directly from the operative, or from poorly cleaned and handled utensils used to dispense the product. For example, ice cream scoops are usually kept in water when not in use, and the growth of microorganisms in this water can cause significant contamination of the scoop, and hence the ice cream (6). Training and personal hygiene of those handling ice cream are therefore very important.

7.4.1   Soft-serve ice cream

Soft-serve ice cream differs from other ice cream products in that it is frozen at the point of sale and does not undergo hardening. The pasteurised mix may be transported to the retail outlet, where it is sold directly from a special dispensing freezer into cones, or onto prepared desserts. Alternatively, the mix may be UHT processed and aseptically packed, or prepared on site from a dried powder blend, or by a conventional process. This system presents a number of opportunities for microbial contamination to occur. Temperature abuse during transport and storage of the unfrozen mix is quite likely, allowing sufficient bacterial growth to cause spoilage. Contamination of mixes during preparation on site is also possible. Inadequate cleaning and sanitation of dispensing freezers may also be a problem, and it is necessary to dismantle and clean such equipment daily. Contamination by *L. monocytogenes* is of particular concern. Some dispensing freezers are now designed to be 'self pasteurising', where all product contact surfaces and residual mixes within the freezer are heated to at least 65 °C for 30 minutes, and then cooled rapidly to 4 °C. A recent UK survey of soft-serve ice cream from fixed and mobile retail outlets showed that there is still cause for concern over the microbiological quality of these products (7).

7.5   Spoilage

Microbiological spoilage will occur only if there is sufficient delay between pasteurisation and freezing. Pasteurisation will destroy most potential spoilage organisms apart from the spores of psychrotrophic bacilli, and microbiological growth does not take place in correctly frozen products. Therefore, the cooling and ageing steps in the process are the most vulnerable for spoilage. This is particularly true if cleaning and sanitation of post-pasteurisation equipment are inadequate, or if flavourings and other ingredients added after pasteurisation are of poor microbiological quality. Therefore, effective control and monitoring of plant hygiene, ingredient quality, and the temperature of mixes between pasteurisation and freezing are vital to prevent spoilage.

The potential for spoilage of soft-serve ice cream mixes has already been mentioned, and it has been shown that even moderate temperature abuse of stored mixes can lead to the development of high bacterial counts and eventual spoilage (8).

7.6   Pathogens: Growth and Survival

Ice creams have a relatively good recent record from a food safety point of view, probably because of the effect of the heat treatment regulations that have been introduced in many countries. Most outbreaks of foodborne disease associated with ice cream have involved ice cream made from raw milk, or home-made products that have used raw milk, cream or eggs, inadequate heat treatment or been contaminated during handling. For example, a recent outbreak of *Salmonella enteritidis* infection involving 30 children, following a birthday party in the UK, was associated with the consumption of home-made ice cream made using raw shell eggs (9). Nevertheless, there have been a number of instances of foodborne disease associated with commercially produced ice cream.

7.6.1   Salmonella

Salmonellae are able to survive for very long periods in ice cream, and, although they will not survive adequate pasteurisation, post-process contamination or the use of raw eggs and failure to pasteurise the ice cream mix, is a serious risk. In 1994, a very large outbreak of *S.enteritidis* infection occurred in Minnesota and other States. The outbreak was estimated to have affected 224,000 people and was associated with a nationally distributed ice cream brand. This was the largest *Salmonella* outbreak ever recorded in the US. The investigation concluded that the probable cause was cross-contamination of pasteurised ice cream mix in tankers also used for transporting unpasteurised raw eggs. The mix was not subsequently repasteurised (10). The infective dose in this outbreak was later calculated as only about 28 cells (11).

7.6.2   Listeria monocytogenes

There has been some concern over the presence of *L. monocytogenes* in ice cream, particularly in view of its ability to grow at low temperatures, and its relatively high heat resistance. It is generally considered that the pasteurisation conditions used in the UK are sufficient to destroy the organism, but that more marginal processes applied elsewhere could be less effective, especially in view of the protective effect of stabilisers mentioned in section 7.3.1.5 . Post-pasteurisation contamination is a potential problem, especially in mixes that are held for long periods prior to freezing. It should be noted that *L. monocytogenes* has been shown to be dispersed in aerosols even at temperatures below 0 °C (12). There have been a number of large recalls of frozen dairy products in the US since 1985, including ice cream bars, vanilla ice milk, and sherbet, because of *Listeria* contamination, although it is not clear whether any of these products caused any cases of illness (13). However, *L. monocytogenes* has been shown to survive freezing and storage in frozen foods for 14 weeks at -18 °C with no reduction in numbers of viable cells (14). Sporadic cases of listeriosis have been reported in Belgium. One notable case was that of a 62 year old immunocompromised man, who consumed ice cream contaminated with *L. monocytogenes* (15).

7.6.3   Staphylococcus aureus

Staph. aureus will not survive ice cream pasteurisation and does not grow at low temperatures. It may be a post-process contaminant introduced via flavourings and other ingredients, or from personnel (via nasal and hand carriers), but is not able to grow and produce enterotoxin unless severe temperature abuse occurs. An outbreak of this type occurred in 1945 at an army hospital in the UK, where a heat-treated mix was cooled slowly overnight before freezing 20 - 30 hours later. Around 700 people were affected (16).

7.6.4   Bacillus cereus

Although there is some concern that psychrotrophic *B. cereus* spores might survive pasteurisation and then grow in the mix during ageing, it seems unlikely that the population would reach sufficient levels to cause illness. However, if the initial number of spores was very high, and time and temperature control after pasteurisation was not adequate, the population could reach high levels, especially in soft-serve mixes. *B. cereus* has been isolated from samples of ice cream (17) and there are reports of outbreaks linked to ice cream (18).

7.6.5   Other pathogens

Food handlers were thought to be responsible for an outbreak of verocytotoxin-producing *Escherichia coli* (VTEC) in 2007. The ice cream, consumed at two birthday parties and at a farm, resulted in five cases of haemolytic uraemic syndrome (HUS) in children, and seven cases of severe diarrhoea (19). These organisms are not heat-resistant and do not grow at low temperatures, but their low infective dose, and their general ability to survive in unfavourable environments suggest that they could pose a serious risk to consumers if inadequate heat treatment or post-pasteurisation contamination occur.

There have also been occasional outbreaks of disease associated with the handling of ice cream during manufacture or at the point of sale. These include a major outbreak of typhoid fever in Wales in 1947, which affected 210 people, with four deaths. The ice cream producer was found to be a urinary excreter of

Salmonella typhi. It was following this outbreak that regulations were introduced in the UK regarding heating of ice cream mixes prior to freezing (18). Outbreaks of paratyphoid, shigella dysentery, and Hepatitis A, as a result of handling by infected individuals, have also been reported, (16, 20). These incidents confirm the importance of health checks and hygiene training for ice cream vendors.

7.7   Toxins

Any risk from mycotoxins in ice cream is likely to be a reflection of ingredient quality. Nuts are the most likely source of aflatoxins, and it is important to ensure that nuts used in ice creams are of high quality, with no evidence of mould growth.

7.8   References

1. Holsinger V.H., Smith P.W., Smith J.L., Palumbo S.A. Thermal destruction of *Listeria monocytogenes* in ice cream mix. *Journal of Food Protection*, 1992, 55 (4), 234-7.

2. Papademas P., Bintsis T.. Microbiology of ice cream and related products, in *Dairy Microbiology Handbook: The Microbiology of Milk and Milk Products*. Ed. Robinson R.. New York, John Wiley. 2002, 213–60.

3. Davidson V.J., Goff H.D., Flores A. Flow characteristics of viscous, non-Newtonian fluids in holding tubes of HTST pasteurisers. *Journal of Food Science*, 1996, 61 (3), 573-6.

4. Alexander J., Rothwell J. A study of some factors affecting the methylene blue test and the effect of freezing on the bacterial content of ice cream. *Journal of Food Technology*, 1970, 5, 387-402.

5. Georgala D.L., Hurst A. The survival of food poisoning bacteria in frozen foods. *Journal of Applied Bacteriology*, 1963, 26 (3), 346-58.

6. Wilson I.G., Heaney J.C.N., Weatherup S.T.C. The effect of ice cream-scoop water on the hygiene of ice cream. *Epidemiology and Infection*, 1997, 119 (1), 35-40.

7. Little C.L., de Louvois J. The microbiological quality of soft ice cream from fixed premises and mobile vendors. *International Journal of Environmental Health Research*, 1999, 9, 223-32.

8. Martin J.H., Blackwood P.W. Effect of pasteurisation conditions, type of bacteria, and storage temperature on the keeping quality of UHT-processed soft-serve frozen dessert mixes. *Journal of Milk and Food Technology*, 1971, 34, 256-9.

9. Dodhia H., Kearney J., Warburton F. A birthday party, home-made ice cream, and an outbreak of *Salmonella enteritidis* phage type 6 infection. *Communicable Disease and Public Health*, 1998, 1 (1), 31-4.

10. Hennessy T.W., Hedberg C.W., Slutsker L., White K.E., Besser-Wiek J.M., Moen M.E., Feldman J., Coleman W.W., Edmonson L.M., MacDonald K.L., Osterholm M.T. A national outbreak of *Salmonella enteritidis* infections from ice cream. *New England Journal of Medicine*, 1996, 334 (20), 1281-6.

11. Vought K.J., Tatini S.R. *Salmonella enteritidis* contamination of ice cream associated with a 1994 multistate outbreak. *Journal of Food Protection*, 1998, 61 (1), 5-10.

12. Goff H.D., Slade P.J. Transmission of a *Listeria* sp. through a cold-air wind tunnel. *Dairy, Food and Environmental Sanitation*, 1990, 10 (6), 340-3.

13. Ryser E.T. Incidence and behaviour of *Listeria monocytogenes* in unfermented dairy products, in *Listeria, Listeriosis and Food Safety*. Eds. Ryser LT., Marth E.H. New York, CRC Press. 2007, 357-403.

14. Palumbo S.A., Williams A.C. Resistance of *Listeria monocytogenes* to freezing in foods. *Food Microbiology*, 1991, 8 (1), 63-8.

15. Andre P., Roose H., Van Noyen R., Dejaegher L., Uyttendaele I., de Schrijver K. Neuro-meningeal listeriosis associated with consumption of an ice cream. *Médecine et Maladies Infectieuses*, 1990, 20, 570-2.

16. Hobbs B.C., Gilbert R.J. *Food Poisoning and Food Hygiene*. London, Arnold. 1978.

17. Ahmed A.A-H., Moustafa M.K., Marth E.H. Incidence of *Bacillus cereus* in milk and some milk products. *Journal of Food Protection*, 1983, 46 (2), 126-8.

18. Griffiths M.W. Milk and unfermented milk products, in *The Microbiological Safety and Quality of Food, Volume 1*. Eds. Lund B.M., Baird-Parker T.C., Gould G.W. Gaithersburg, Aspen Publishers. 2000, 507-34.

19. De Schrijver K., Possé B., Van den Branden D., Oosterlynck O., De Zutter L., Eilers K., Piérard D., Dierick K., Van Damme-Lombaerts R., Lauwers C., Jacobs R. Outbreak of verocytotoxin-producing *E.coli* O145 and O26 infections associated with the consumption of ice cream produced at a farm, Belgium, 2007. *Eurosurveillance*, 2008, 13 (7), 8041.

20. MacDonald K.L., Griffin P.M. Foodborne disease outbreaks, annual summary, 1982. *Morbidity and Mortality Weekly Report*, 1983, 35, 7.

8. HACCP

8.1 Introduction

The Hazard Analysis Critical Control Point (HACCP) system is a structured, preventative approach to ensuring food safety. HACCP provides a means to identify and assess potential hazards in food production and establish preventive control procedures for those hazards. A critical control point (CCP) is identified for each significant hazard, where effective control measures can be defined, applied and monitored. The emphasis on prevention of hazards reduces reliance on traditional inspection and quality control procedures and end-product testing. A properly applied HACCP system is now internationally recognised as an effective means of ensuring food safety.

The HACCP concept can be applied to new or existing products and processes, and throughout the food chain from primary production to consumption. It is compatible with existing standards for quality management systems such as the ISO 9000-2000 series, and HACCP procedures can be fully integrated into such systems. The new ISO 22000 food safety standard formally integrates HACCP within the structure of a quality management system. HACCP is fully integrated into the British Retail Consortium (BRC) Global Standards for Food Safety, and is one of the 'fundamental' requirements of that system.

The application of HACCP at all stages of the food supply chain is actively encouraged, and increasingly required, worldwide. For example, the Codex Alimentarius advises that "the application of HACCP systems can aid inspection by regulatory authorities and promote international trade by increasing confidence in food safety".

In many countries, there is a legal requirement for all food business operators to have some form of hazard analysis based on HACCP as a means of ensuring food safety. For example, within the European Union, Regulations 852/2004 and 853/2004 require a fully operational and maintained HACCP system, according to Codex, to be in place.

8.2 Definitions

Control (verb) - To take all necessary actions to ensure and maintain compliance with criteria established in the HACCP plan.

Control (noun) - The state wherein correct procedures are followed and criteria are met.

Control measure - An action and activity that can be used to prevent or eliminate a food safety hazard or reduce it to an acceptable level.

Corrective action - An action to be taken when the results of monitoring at the CCP indicate a loss of control.

Critical Control Point (CCP) - A step at which control can be applied and is essential to prevent or eliminate a food safety hazard, or reduce it to an acceptable level.

Critical limit - A criterion that separates acceptability from unacceptability.

Deviation - Failure to meet a critical limit.

Flow diagram – A systematic representation of the sequence of steps or operations used in the production or manufacture of a particular food item.

HACCP - A system that identifies, evaluates and controls hazards that are significant for food safety.

HACCP Plan – A document prepared in accordance with the principles of HACCP to ensure control of hazards that are significant for safety in the segment of the food chain under consideration.

Hazard - A biological, chemical or physical agent in, or condition of, food with the potential to cause an adverse health effect.

Hazard analysis - The process of collecting and evaluating information on hazards and the conditions leading to their presence to decide which are significant for food safety and therefore should be addressed by the HACCP plan.

Monitoring – The act of conducting a planned sequence of observations or measurements of control parameters to assess whether a CCP is under control.

Step - A point, procedure, operation or stage in the food chain including raw materials, from primary production to final consumption.

Validation - Obtaining evidence that the elements of the HACCP plan are effective.

Verification - The application of methods, procedures, tests and other evaluations, in addition to monitoring to determine compliance with the HACCP plan.

8.3   Stages of a HACCP Study

The HACCP system consists of the following seven basic principles:

1. Conduct a hazard analysis.

2. Identify the CCPs.

3. Establish the critical limit(s).

4. Establish a system to monitor control of the CCP.

5. Establish the corrective action to be taken when monitoring indicates that a particular CCP is not under control.

6. Establish procedure for verification to confirm that the HACCP system is working effectively.

7. Establish documentation concerning all procedures and records appropriate to these principles and their application.

It is recommended by the Codex Alimentarius that the practical application of the HACCP principles be approached by breaking the seven principles down into a 12-stage logic sequence. Each stage is discussed below in detail. Figure 8.1 is a flow diagram illustrating this 12-stage logic sequence.

8.3.1   Assemble the HACCP team

HACCP requires management commitment of resources to the process. An effective HACCP plan is best carried out as a multidisciplinary team exercise to ensure that the appropriate product-specific expertise is available. The team should include members familiar with all aspects of the production process as well as specialists with expertise in particular areas such as production, hygiene managers, quality assurance or control, ingredient and packaging buyers, food microbiology, food chemistry or engineering. The team should also include personnel who are involved with the variability and limitations of the operations. If expert advice is not available on-site, it may be obtained from external sources.

The scope of the plan should be determined by defining the extent of the production process to be considered and the categories of hazard to be addressed (e.g. biological, chemical and/or physical).

8.3.1.1 Dairy products

The HACCP team should ideally have access to expertise on the practices applied at farm level in relation to milk collection, storage and transport. The initial microbial population of raw milk has a significant influence on the safety and quality of processed dairy products. For example, the effectiveness of pasteurisation may be compromised by excessive microbial counts in raw milk, and by the presence of large numbers of pathogens. Therefore, knowledge of primary production procedures is very valuable for the HACCP study.

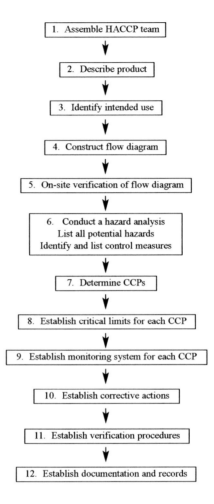

Fig. 8.1. Logic sequence for application of HACCP

8.3.2 Describe the product

It is important to have a complete understanding of the product, which should be described in detail. The description should include information such as the product name, composition, physical and chemical structure (including water activity (a_w), pH, etc.), processing conditions (e.g. heat treatment, freezing, fermentation, etc.), packaging, shelf life, storage and distribution conditions and instructions for use.

8.3.2.1 Dairy products

Many dairy products are manufactured by traditional processes that have been practised for centuries. As a result of this, there is a great deal of background data and experience available to draw on. Furthermore, the majority of these traditional products have a good safety record, suggesting that standard manufacturing processes are safe. This situation can lead to complacency, and it is essential that the basis for the inherent safety of these products is fully understood. This is particularly true in situations where the introduction of new technology, new additives and ingredients, and new requirements from retailers and consumers may give rise to new hazards.

8.3.3 Identify intended use

The intended use should be based on the expected uses of the product by the end-user or consumer (e.g. is a cooking process required?). It is also important to identify the consumer target groups. Vulnerable groups of the population, such as children or the elderly, may need to be considered specifically.

8.3.3.1 Dairy products

Dairy products are often consumed by high-risk groups, particularly the very young and the elderly. Infants are at particular risk from pathogens such as *Salmonella*, and pregnant women and the elderly are especially vulnerable to *Listeria* infection. This must be considered during the HACCP study and should be taken into account when compiling the instructions for use.

8.3.4 Construct a flow diagram

The flow diagram should be constructed by the HACCP team and should contain sufficient technical data for the study to progress. It should provide an accurate representation of all steps in the production process from raw materials to the end-product. It may include details of the factory and equipment layout, ingredient specifications, features of equipment design, time/temperature data, cleaning and hygiene procedures and storage conditions. Ideally it should also include details of CCP steps, once determined.

8.3.4.1 Dairy products

Examples of flow diagrams for specific dairy products may be found in the appropriate product chapters. Many dairy processing operations have relatively few steps and the flow diagrams appear simple. Common steps occur in many processes - for example, standardisation, pasteurisation, and homogenisation.

However, it is essential that the details of each step are fully appreciated and recorded. Particular attention should be paid to potential routes of product flow that might allow cross-contamination between raw and pasteurised product. Divert valves, bypasses, pumps, and holding or balance tanks require close scrutiny. In modern dairy plants, it is also important to ascertain how cleaning-in-place systems are designed and operated. Effective cleaning is an essential control for preventing recontamination of pasteurised dairy products.

8.3.5   On-site confirmation of the flow diagram

The HACCP team should confirm that the flow diagram matches the process that is actually being carried out. The operation should be observed at all stages, and any discrepancies between the flow diagram and normal practice must be recorded and the diagram amended accordingly. It is also important to include observation of production outside normal working hours such as shift patterns and weekend working, as well as the circumstances of any reclaim or rework activity. It is essential that the diagram is accurate, because the hazard analysis and decisions regarding CCPs are based on these data. If HACCP studies are applied to proposed new process lines/ products, then any pre-drawn HACCP plans must be reviewed once the lines/products are finalised.

8.3.6   List all potential hazards associated with each step; conduct a hazard analysis; and identify any measures to control identified hazards

The HACCP team should list all hazards that may reasonably be expected to occur at each step in the production process.

The team should then conduct a hazard analysis to identify which hazards are of such a nature that their elimination or reduction to an acceptable level is essential to the production of safe food.

The analysis is likely to include consideration of:

- The likely occurrence of hazards and the severity of their adverse health effects;

- The qualitative and/or quantitative evaluation of the presence of hazards;

- Survival or multiplication of pathogenic microorganisms;

- Production or persistence of toxins;

- The hurdle effect;

- The number of consumers potentially exposed and their vulnerability;

- Any food safety objectives or manufacturer's food safety requirements.

The HACCP team should then determine what control measures exist that can be applied for each hazard.

Some hazards may require more than one control measure for adequate control and a single control measure may act to control more than one hazard. One control measure may be relevant to several process steps, where a hazard is repeated.

Note: it is important at this stage that no attempt is made to identify CCPs, since this may interfere with the analysis.

8.3.6.1 Dairy products

The term 'dairy products' includes a varied group of foods, and there is an equally varied range of potential hazards associated with them. Hazards specific to certain types of product are detailed in the appropriate chapters of this manual. For example, there are particular hazards associated with contamination of dried milk powders by salmonellae, and the potential growth of *Listeria monocytogenes* in soft cheeses.

Many of the microbiological hazards associated with dairy products are derived from the raw materials (i.e. raw milk). Pathogens may be part of the resident microflora of the Iiving animal (e.g. *Staphylococcus aureus*), or may originate from faecal contamination during initial milk collection (e.g. *Salmonella* and *E.coli* 0157). Pathogens may also be introduced into raw milk from contaminated equipment during collection, transport, or storage. The majority of these hazards can be eliminated by an appropriate heat treatment, such as pasteurisation or sterilisation.

Hazards introduced during processing of dairy products depend very much on the characteristics of the process. For example, heat-sensitive pathogens may be present in pasteurised milk as a result of cross-contamination between raw and heat-treated milk, and slow acid production by the starter culture in fermented milk products may allow growth and toxin production by *Staph. aureus*. Therefore, it is not possible, or desirable, to generalise about expected hazards, and the reader is referred to the appropriate product chapter in this book for additional advice on specific hazards.

8.3.7 Determine CCPs

The determination of CCPs in the HACCP system is facilitated by using a decision tree (Figure 8.2) to provide a logical, structured approach to decision making. However, application of the decision tree should be flexible, and its use may not always be appropriate. It is also essential that the HACCP team has access to sufficient technical data to determine the CCPs effectively.

If a significant hazard has been identified at a step where control is required for safety, but for which no control exists at that step or any other, then the process must be modified to include a control measure.

Answer the following questions for each identified hazard:

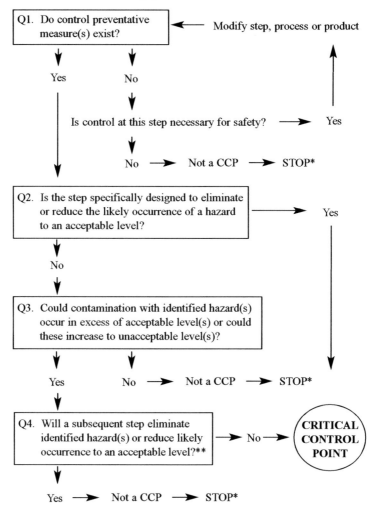

*   Proceed to next step in the described process
** Acceptable and unacceptable levels need to be defined within the
    overall objectives in identifying the CCPs of HACCP plan

Fig. 8.2. CCP Decision Tree A
(Adapted from Codex Alimentarius Commission, 1997)

8.3.7.1 Dairy products

Again, given the enormous variety of dairy products and processes in use, it is unwise to generalise on likely CCPs, and the reader is referred to the appropriate product chapter in this book. However, it can be said that effective control measures are likely to include the following:

- Careful control of raw milk quality and selection of sources for other raw materials;

- Adequate pasteurisation processes;

- Prevention of cross-contamination of pasteurised product;

- Effective sanitation and hygiene procedures;

- Adequate temperature control.

Some examples are as follows:

In the manufacture of skimmed milk powder, CCPs are likely to be pasteurisation, and the effective separation, cleaning and maintenance of spray dryers and powder handling equipment.

In the production of fermented milk products and cheese, pasteurisation is again likely to be a CCP, but the rapid development of sufficient acidity by the starter culture is also a CCP.

Adequate temperature control during processing would normally be considered a CCP in the manufacture of ice cream, as would the microbiological quality of flavouring ingredients added after pasteurisation.

8.3.8   Establish critical limits for each CCP

Critical limits separate acceptable from unacceptable products. Where possible, critical limits should be specified and validated for each CCP. More than one critical limit may be defined for a single step. For example, it is necessary to specify both time and temperature for a thermal process, and a minimum process of 72 °C for 15 seconds, or equivalent, is required for milk pasteurisation. Criteria used to set critical limits must be measurable and may include physical, chemical, biological or sensory parameters.

It is prudent to set stricter limits (often called target or process limits/levels) to ensure that any trends towards a loss of control is noted before the critical limit is exceeded.

8.3.8.1 Dairy products

Specific product chapters provide information on criteria that may be used to set critical limits. Some examples relevant to dairy products are:

- Pasteurisation time and temperature
- Total acidity and/or pH (fermented products)
- Measured adequacy of cleaning procedures
- Chilled storage time and temperature
- Water activity (condensed milk products)

8.3.9 Establish a monitoring system for each CCP

Monitoring involves planned measurement or observation of a CCP relative to its critical limits. Monitoring procedures must be able to detect loss of control of the CCP, and should provide this information with sufficient speed to allow adjustments to be made to the control of the process before the critical limits are violated. Monitoring at critical limits should be able to detect rapidly when the critical limit has been exceeded. Monitoring should either be continuous, or carried out sufficiently frequently to ensure control at the CCP. Therefore, physical and chemical on-line measurements are usually preferred to lengthy microbiological testing. However, certain rapid methods, such as ATP assay by bioluminescence, may be useful for assessment of adequate cleaning, which could be a critical limit for some CCPs, for example, pre-start-up hygiene.

Persons engaged in monitoring activities must have sufficient knowledge, training and authority to act effectively on the basis of the data collected. These data should also be properly recorded.

8.3.10 Establish corrective actions

For each CCP in the HACCP plan, there must be specified corrective actions to be applied if the CCP is not under control. If monitoring indicates a deviation from the critical limit for a CCP, action must be taken that will bring it back under control. Actions taken should also include proper isolation of the affected product and an investigation into why the deviation occurred. A further set of corrective actions should relate to the target level, if process drift is occuring. In this case, only repair of the process defect and investigation of the fault are required. All corrective actions should be properly recorded.

8.3.11 Establish verification procedures

Verification usually involves auditing and testing procedures. Auditing methods, procedures and tests should be used frequently enough to determine whether the HACCP system is being followed, and is effective at controlling the hazards. These may include random sampling and analysis, including microbiological testing. Although microbiological analysis is generally too slow for monitoring purposes, it can be of great value in verification, since many of the identified hazards are likely to be microbiological. For example, analysis of dried milk powders for *Salmonella*, desserts for *Bacillus cereus*, and soft cheeses for *Listeria* would be appropriate verification tests.

In addition, reviews of HACCP records are important for verification purposes. These should confirm that CCPs are under control and should indicate the nature of any deviations and the actions that were taken in each case. It is also useful to review customer returns and complaints regularly.

8.3.12 Establish documentation and record keeping

Efficient and accurate record keeping is an essential element of a HACCP system. The procedures in the HACCP system should be documented.

Examples of documented procedures include:

- The hazard analysis

- Determination of CCPs

- Determination of critical limits

- The completed HACCP plan

Examples of recorded data include:

- Results of monitoring procedures

- Deviations from critical limits and corrective actions

- Records of certain verification activities, e.g. observations of monitoring activities, and calibration of equipment.

The degree of documentation required will depend partly on the size and complexity of the operation, but it is unlikely to be possible to demonstrate that an effective HACCP system is present without adequate documentation and records. The length of time that records are kept will be as per company policy, but should not be less than one year beyond the shelf life of the product. Three to five years is typical for many food companies.

8.4   Implementation and Review of the HACCP Plan

The completed plan can only be implemented successfully with the full support and co-operation of management and the workforce. Adequate training is essential and the responsibilities and tasks of the operating personnel at each CCP must be clearly defined.

Finally, it is essential that the HACCP plan be reviewed following any changes to the process, including changes to raw materials, processing conditions or equipment, packaging, cleaning procedures and any other factor that may have an effect on product safety. Even a small alteration to the product or process may invalidate the HACCP plan and introduce potential hazards. Therefore, the implications of any changes to the overall HACCP system must be fully considered and documented, and adjustments made to the procedures as necessary.

Tiggered reviews/audits should occur as a result of changes, whereas scheduled review/audit should be annually, as a minimum.

8.5   References

Wareing, P.W., Carnell, A.C. *HACCP – A Toolkit for Implementation*. Leatherhead, Leatherhead Food International. 2007.

Drosinos E.H., Siana P.S. HACCP in the cheese manufacturing process, a case study, in *Food Safety: A Practical and Case Study Approach*. Eds. McElhatton A., Marshall R.J. Berlin, Springer. 2007, 90-111.

Bernard D., Scott. V. Hazard Analysis and Critical Control Point System: use in controlling microbiological hazards, in *Food Microbiology: Fundamentals and Frontiers*. Eds. Doyle M.P., Beuchat L.R. Washington DC, ASM Press. 2007, 971-86.

Rabi A., Banat A., Shaker R.R., Ibrahim S.A. Implementation of HACCP system to large scale processing line of plain set yogurt. *Italian Food and Beverage Technology*, 2004, (35), 12-17.

Institute of Medicine, National Research Council. Scientific criteria and performance standards to control hazards in dairy products, in *Scientific Criteria to Ensure Safe Food*. Ed. Institute of Medicine, National Research Council. Washington D.C., National Academic Press. 2003, 225-47.

Jervis D. Application of Process Control, in *Dairy Microbiology Handbook: The Microbiology of Milk and Milk Products*. Ed. Robinson R. New York, John Wiley and Sons, Inc. 2002, 593-654.

Mortimore S., Mayes T. The effective implementation of HACCP systems in food processing, in *Foodborne Pathogens: Hazards, Risk Analysis and Control*. Eds. Blackburn C. de W., McClure P.J. Cambridge, Woodhead Publishing Ltd. 2002, 229-56.

Ali A.A., Fischer R.M. Implementation of HACCP to bulk condensed milk production line. *Food Reviews International*, 2002, 18 (2-3), 177-90.

Mayes T., Mortimore C.A. *Making the most of HACCP: Learning from Other's Experience.* Cambridge, Woodhead Publishing. 2001.

Mortimore S.E., Wallace C., Cassianos C. *HACCP (Executive Briefing).* London, Blackwell Science Ltd. 2001.

Dairy Practices Council. Hazard Analysis Critical Control Point system - HACCP for the dairy industry. Guideline No. 55, in *Guidelines for the Dairy Industry Relating to Sanitation and Milk Quality, Volume 4.* Ed. Dairy Practices Council. Keyport, DPC. 2001.

Sandrou D.K., Arvanitoyannis I.S. Application of Hazard Analysis Critical Control Point (HACCP) system to the cheese making industry: a review. *Food Reviews International*, 2000, 16 (3), 327-68.

Sandrou D. K., Arvanitoyannis I.S. Implementation of Hazard Analysis Critical Control Point (HACCP) to the dairy industry: current status and perspectives. *Food Reviews International*, 2000, 16 (1), 77-111.

Gould B.W., Smukowski M., Bishop J.R. HACCP and the dairy industry: an overview of international and US experiences, in *The Economics of HACCP: Costs and Benefits.* Ed. Unnevehr L.J. St Paul, Eagan Press. 2000, 365-84.

Chartered Institute of Environmental Health. *HACCP in Practice.* London, Chadwick House Group Ltd. 2000.

Jouve J.L. Good manufacturing practice, HACCP, and quality systems, in *The Microbiological Safety and Quality of Food, Volume 2.* Eds. Lund B.M., Baird-Parker T.C., Gould G.W. Gaithersburg, Aspen Publishers. 2000, 1627-55.

Stevenson K.E., Bernard D.T. *HACCP: A Systematic Approach to Food Safety.* Washington DC, Food Processors Institute. 1999.

Mavropoulos A.A., Arvanitoyannis I.S. Implementation of Hazard Analysis Critical Control Point to Feta and Manouri cheese production lines. *Food Control*, 1999, 10 (3), 213-9.

Corlett D.A. *HACCP User's Manual.* Gaithersburg, Aspen Publishers. 1998.

Mortimore S., Wallace C. *HACCP: A Practical Approach.* Gaithersburg, Aspen Publishers. 1998.

Khandke S.S., Mayes T. HACCP implementation: a practical guide to the implementation of the HACCP plan. *Food Control*, 1998, 9 (2-3), 103-9.

Forsythe S.J., Hayes P.R. *Food Hygiene, Microbiology and HACCP.* Gaithersburg. Aspen Publishers. 1998.

Food and Agriculture Organisation. *Food Quality and Safety Systems: A Training Manual on Food Hygiene and the Hazard Analysis and Critical Control Point (HACCP) system.* Rome, FAO. 1998.

National Advisory Committee on Microbiological Criteria for Foods. *Hazard Analysis and Critical Control Point Principles and Application Guidelines.* 1997.

Gardner L.A. Testing to fulfil HACCP (Hazard Analysis Critical Control Points) requirements: principles and examples. *Journal of Dairy Science*, 1997, 80 (12), 3453-7.

Codex Alimentarius Commission. Hazard Analysis Critical Control Point (HACCP) System and guidelines for its application, in *Food Hygiene: Basic texts*. Ed. Codex Alimentarius Commission. Rome, FAO. 1997, 33-45.

Savage R.A. Hazard Analysis Critical Control Point: a review. *Food Reviews International*, 1995, 11 (4), 575-95.

Peta C, Kailasapathy K. HACCP - its role in dairy factories and the tangible benefits gained through its implementation. *Australian Journal of Dairy Technology*, 1995, 50 (2), 74-8.

Pierson M.D., Corlett D.A., Institute of Food Technologists. *HACCP: Principles and Applications*. New York, Van Nostrand Reinhold. 1992.

Bryan F.L., World Health Organisation. *Hazard Analysis Critical Control Point Evaluations: A Guide to Identifying Hazards and Assessing Risks Associated with Food Preparation and Storage*. Geneva, WHO. 1992.

Mayes T. Simple users' guide to the hazard analysis critical control point concept for the control of food microbiological safety. *Food Control*, 1992, 3 (1), 14-19.

International Commission on Microbiological Specifications for Foods. *Microorganisms in Foods, Volume 4: Application of the Hazard Analysis Critical Control Point (HACCP) System to Ensure Microbiological Safety and Quality*. Oxford, Blackwell Scientific Publications. 1988.

9. EU FOOD HYGIENE LEGISLATION

9.1   Introduction

Hygiene is an important aspect of ensuring food safety and one that plays an important role in most countries' food legislation. Hygiene is a general concept that covers a wide subject area, from structural conditions in the factory or process facility, to personnel requirements, final product specifications, including microbiological criteria, transport and delivery vehicles requirements, and conditions of raw materials.

Microbiological standards have a useful role and help establish requirements for the microbiological safety and quality of food and raw materials. A number of standards are provided in food legislation; however, the existence of microbiological standards cannot protect consumer health alone. It is generally considered that the principles of Good Manufacturing Practice (GMP) and application of Hazard Analysis Critical Control Point (HACCP) systems are of greater importance.

A new package of EU hygiene measures became applicable on 1 January 2006 to update and consolidate the earlier 17 hygiene directives with the intention of introducing consistency and clarity throughout the food production chain from primary production to sale or supply to the final consumer. The general food hygiene Directive 93/43/EEC and other Directives on the hygiene of foodstuffs and the health conditions for the production and placing on the market of certain products of animal origin intended for human consumption have been replaced by several linked measures on food safety rules and associated animal health controls.

The new legislation was designed to establish conditions under which food is produced to optimise public health and to prevent, eliminate or acceptably control pathogen contamination of food. Procedures under the new legislation are based on risk assessment and management and follow a 'farm to fork' approach to food safety with the inclusion of primary production in food hygiene legislation. Prescribed are detailed measures to ensure the safety and wholesomeness of food during preparation, processing, manufacturing, packaging, storing, transportation, distribution, handling and offering for sale or supply to the consumer.

9.2 Legislative Structure

From 1 January 2006, the following EU hygiene regulations have applied:

- Regulation (EC) No. 852/2004 of the European Parliament and of the Council on the hygiene of foodstuffs

- Regulation (EC) No. 853/2004 of the European Parliament and of the Council laying down specific hygiene rules for food of animal origin

- Regulation (EC) No. 854/2004 of the European Parliament and of the Council laying down specific rules for the organisation of official controls on products of animal origin intended for human consumption

- Regulation (EC) No. 2073/2005 on microbiological criteria for foodstuffs

The general hygiene requirements for all food business operators are laid down in Regulation 852/2004. Regulation 853/2004 supplements Regulation 852/2004 in that it lays down specific requirements for food businesses dealing with foods of animal origin. Regulation 854/2004 relates to the organisation of official controls on products of animal origin and sets out what those enforcing the provisions have to do.

N.B. A number of more detailed implementing and transitional measures have been adopted at EC level.

Subsequently, existing hygiene Directives including those below were repealed:

- Commission Directive 89/362/EEC of 26 May 1989 on general conditions of hygiene in milk production holdings OJ L 156, 8.6.1989, 30–2

- Council Directive 92/46/EEC of 16 June 1992 laying down the health rules for the production and placing on the market of raw milk, heat-treated milk and milk-based products OJ L 268, 14.9.1992, 1–32

- Council Directive 93/43/EEC of 14 June 1993 on the hygiene of foodstuffs OJ L 175, 19.7.1993, 1–11

The EU hygiene regulations apply to all stages of food production including primary production.

As regulations, the legislation is directly applicable law and binding in its entirety on all member states from the date of entry into force.

Although the regulations have the force of law, national legislation in the form of a Statutory Instrument (S.I.) in England, and equivalent legislation in Scotland, Wales and Northern Ireland, is required to give effect to the EU regulations, for example, setting offences, penalties and powers of entry, revocation of existing implementing legislation, etc.

The Food Hygiene (England) Regulations 2006 (S.I. 2006 No.14, as amended) came into force on 11 January 2006 (separate but similar national legislation also came into force on that day in Scotland, Wales and Northern Ireland). The national legislation in all four UK countries also applied the provisions of the EU Microbiological Criteria Regulation No. 2073/2005.

Although EU food hygiene regulations are directly applicable in the individual Member States there are some aspects where Member States are required or allowed to adopt certain provisions into their national laws. In England for example, there are temperature requirements for foods laid down in Schedule 4 of the Food Hygiene (England) Regulations 2006, as amended. Also, for England, there are restrictions on the sale of raw milk intended for human consumption as laid down by schedule 6 of the Food Hygiene (England) Regulations 2006, as amended. Both of these issues will be covered later in this chapter.

9.3   Regulation (EC) No. 852/2004 on the General Hygiene of Foodstuffs

Food business operators must ensure that all stages of production, processing and distribution of food under their control satisfy the relevant hygiene requirements laid down in Regulation (EC) No. 852/2004.

This Regulation lays down general rules for food business operators on the hygiene of foodstuffs, particularly taking into account a number of factors ranging from ensuring food safety throughout the food chain to begin with primary production, right through to the implementation of procedures based on HACCP principles.

There are some exemptions, for example, with primary production, domestic preparation or handling, food storage that is for private or domestic consumption, and also if the producer supplies small amounts of primary product to the final consumer or local retail establishments supplying the final consumer, since Regulation (EC) 852/2004 will not apply in these cases. Likewise Regulation 852/2004 will not apply to collection centres and tanneries meeting the definition of food business because they handle raw material for the production of gelatine or collagen.

The regulation lays down general hygiene provisions for which food business operators carrying out primary production must comply with as laid down in Part A of Annex I. Additionally the requirements of EC regulation 853/2004 must be complied with which will be covered later in this chapter.

9.3.1   *Annex I - Primary Production*

Annex I (Part A) relates to general hygiene provisions for primary production and associated operations covering:

(a) the transport, storage and handling of primary products at the place of production

(b) the transport of live animals

(c) for products of plant origin - transport operations to deliver primary products (which haven't been substantially altered) from the place of production to an establishment

Food business operators have the responsibility to ensure primary products are protected against contamination. Any community and national legislation relating to the control of hazards in primary production such as measures to control contamination resulting from surroundings, for example, air, soil, water etc. and measures relating to animal health and welfare, and plant health that may impact on human health should be complied with. Food business operators rearing, harvesting or hunting animals or producing primary products of animal origin are to take adequate measures as necessary. Therefore this relates to the cleaning and disinfection of equipment, and the storage and handling of waste.

Requirements for record keeping are also laid down. This relates to animal feed (nature and origin), veterinary medicines administered to animals (date given and withdrawal periods), any diseases, analysis of samples from other animals which might impact on human health as well as reports on animal checks performed.

9.3.2   Annex II - Stages Other Than Primary Production

Annex II of the regulation lays down additional general hygiene requirements that must be met by food business operators carrying out production, processing and distribution of food following those stages above. A summary of Chapters I to IV of Annex II is provided in the following sections.

9.3.2.1   Chapter I

Chapter I applies to all food premises, except premises to which Chapter III applies.
- Food premises must be kept clean and maintained in good repair and condition. The layout should allow for this.
- The environment should allow good hygiene practices and give temperature controlled handling and storage conditions where necessary, and to allow foods to be kept at correct temperatures and be monitored.
- Additionally there are requirements for adequate lavatories, basins, ventilation, lighting and draining.

9.3.2.2   Chapter II

Chapter II applies to all rooms where food is prepared, treated or processed, except dining areas and premises to which Chapter III applies.
- The design and layout of rooms should allow for good hygiene practices between and during operations. Therefore floor and wall surfaces, ceilings and windows should be constructed to prevent dirt accumulating.

- Surfaces where food is handled must be maintained well and allow easy cleaning and disinfection preferably using smooth, washable corrosion-resistant and non-toxic materials.

- There should be facilities for cleaning or disinfecting, and for storing working utensils or equipment. Clean potable water and adequate provision for washing food is needed.

9.3.2.3 Chapter III

Chapter III applies to temporary premises (e.g. marquees, market stalls, mobile sales vehicles), premises used primarily as a private dwelling-house but where foods are regularly prepared for placing on the market, and vending machines.

- Here, premises and vending machines should practically be sited, designed, constructed and kept clean and maintained in good repair and condition so as to avoid the risk of contamination, in particular by animals and pests.

- Facilities should allow adequate personal hygiene and surfaces in contact with food should be easy to clean. Enough potable water and storage arrangements for hazardous or inedible substances is required as well as adherence to food safety requirements.

9.3.2.4 Chapter IV

Chapter IV applies to all transportation.

- This lists requirements that conveyances and/or containers used for transporting foodstuffs are to be kept clean and maintained in good repair and condition to protect foodstuffs from contamination and are, where necessary, to be designed and constructed to permit adequate cleaning and/or disinfection.

- Food should be maintained at appropriate temperatures.

9.3.2.5 Chapter V

Chapter V refers to equipment requirements.

- Adequate cleaning and disinfection is to be done frequently for articles, fittings and equipment contacting food where contamination needs to be avoided.

- Equipment should be installed to allow adequate cleaning, and be fitted with the required control device.

9.3.2.6 Chapter VI

Chapter VI refers to food waste.

- Food waste, non-edible by-products and other refuse is to be removed from rooms where food is present as quickly as possible to avoid accumulation. Such waste is to be deposited in closable containers, to allow easy cleaning.

- Refuse stores should allow easy cleaning and be free of pests.
- Waste must be eliminated hygienically in accordance with community legislation.

9.3.2.7 *Chapter VII*

Chapter VII refers to water supply.
- There are requirements that there should be an adequate supply of potable water, requirements for recycled water, ice contacting food, steam used and for the water used in the cooling process for heat treated foods in hermetically sealed containers.

9.3.2.8 *Chapter VIII*

Chapter VIII is about personal hygiene required by those working in a food handling area including clean protective clothing and that those carrying or suffering from a disease are not permitted to handle food.

9.3.2.9 *Chapter IX*

Chapter IX covers provisions applicable to foodstuffs.
- A food business operator should not accept raw materials or ingredients, other than live animals, or any other material used in processing products, if they are known to be contaminated with parasites, pathogenic microorganisms or foreign substances. Neither should they accept raw materials or ingredients that are toxic or decomposed to such an extent that, even after the business operator applied normal processing hygienically, the product would be inedible.
- Raw materials must be kept under appropriate conditions throughout production, processing and distribution. In particular, temperature control (i.e. cold chain and food thawing) requirements are laid down.

9.3.2.10 *Chapter X*

Chapter X lays down provisions applicable to the wrapping and packaging of foodstuffs to avoid contamination of any form.

9.3.2.11 *Chapter XI*

Chapter XI lays down heat treatment requirements for food that is placed on the market in hermetically sealed containers.
- The process used should comply with internationally recognised standards (i.e. pasteurisation, Ultra High Treatment or sterilisation)

9.3.2.12 Chapter XII

Chapter XII states training requirements for food business operators to ensure that food handlers are trained in food hygiene matters and in the application of HACCP principles.

9.3.3   Registration

The regulation requires that food business operators must notify their competent authority of their establishment and have it registered.

Food business operators must also ensure that the competent authority always has up-to-date information on establishments, including the notification of significant changes in activity and closure of an existing establishment.

Food business operators must ensure that establishments are approved by the competent authority, following at least one on-site visit, when approval is required by the national law of the Member State, or under Regulation (EC) No. 853/2004, or by a separate decision adopted.

Separate rules apply for businesses producing products of animal origin.

9.3.4   HACCP

Food business operators, other than at the level of primary production, and associated operations must put in place, implement and maintain a permanent procedure or procedures based on principles of the system of hazard analysis and critical control points (HACCP). Emphasis is placed on risk-related control, with responsibility placed on the proprietor of the food business to ensure that potential hazards are identified and systems are developed to control them. Under HACCP, food business operators must, amongst others, identify hazards to be prevented, eliminated or reduced to acceptable levels, identify and establish critical control points (CCP) to prevent, eliminate or reduce hazards to allow this to be monitored, and establish corrective actions in the case where a CCP is out of control. Procedures must be taken to confirm the above is in place and up-to-date, as well as provide documents and records as evidence for the competent authority.

9.4   Regulation (EC) No. 853/2004 Laying Down Specific Hygiene Rules for Food of Animal Origin

Regulation (EC) No. 853/2004 lays down hygiene rules for products of animal origin which apply in addition to the general hygiene rules of Regulation (EC) No. 852/2004.

9.4.1   Definitions

Dairy products are processed products resulting from the processing of raw milk or from the further processing of such processed products.

Raw milk is milk produced by the secretion of the mammary gland of farmed animals that has not been heated to more than 40 °C or undergone any treatment that has an equivalent effect.

Milk production holding means an establishment where one or more farm animals are kept to produce milk with a view to placing it on the market as food.

Colostrum is the fluid secreted by the mammary glands of milk-producing animals up to three to five days post parturition, that is rich in antibodies and minerals and precedes the production of raw milk

Colostrum-based products are processed products resulting from the processing of colostrum or from the further processing of such processed products.

9.4.2   Requirements

The regulation details specific hygiene requirements for raw milk, colostrum, dairy products and colostrum-based products. Extracts of the requirements of regulation 853/2004, as amended, specifically relating to milk and milk products are given below; for full requirements, reference should be made to the actual regulation.

Food business operators producing or, as appropriate, collecting raw milk and colostrum must ensure compliance with the requirements laid down in Annex III, Section IX as follows:

9.4.2.1   Chapter I: Raw Milk – Primary Production

9.4.2.1.1 Health requirements for raw milk and colostrum production

Raw milk and colostrum must come from animals free from any symptoms of infectious diseases that can be transferred to humans though milk and colostrum. Therefore such milk and colostrum needs to come from:

(i)   cows or buffaloes belonging to a herd which, within the meaning of Directive 64/432/EEC, is free or officially free of brucellosis;

(ii) sheep or goats belonging to a holding free or officially free of brucellosis within the meaning of Directive 91/68/EEC; or

(iii) females of other species, for species susceptible to brucellosis, belonging to herds regularly checked for that disease under a control plan that the competent authority has approved.

Likewise, the same conditions apply in relation to tuberculosis.

There are cases whereby raw milk from animals that do not meet the requirements of the above may be used with the authorisation of the competent authority such as:

(a) in the case of cows or buffaloes that do not show a positive reaction to tests for tuberculosis or brucellosis, nor any symptoms of these diseases, after having undergone a heat treatment such as to show a negative reaction to the alkaline phosphatase test;

(b) in the case of sheep or goats that do not show a positive reaction to tests for brucellosis, or which have been vaccinated against brucellosis as part of an approved eradication programme, and which do not show any symptom of that disease, either:

(i) for the manufacture of cheese with a maturation period of at least two months; or

(ii) after having undergone heat treatment such as to show a negative reaction to the alkaline phosphatase test; and

(c) in the case of females of other species that do not show a positive reaction to tests for tuberculosis or brucellosis, nor any symptoms of these diseases, but belong to a herd where brucellosis or tuberculosis has been detected after the following, checks provided it is treated to ensure its safety:

- females of other species belonging, for species susceptible to brucellosis, to herds regularly checked for that disease under a control plan that the competent authority has approved.

- females of other species belonging, for species susceptible to tuberculosis, to herds regularly checked for this disease under a control plan that the competent authority has approved.

9.4.2.1.2 Hygiene on milk and colostrum production holdings

A. Requirements for premises and equipment
This relates to milking equipment and premises where milk and colostrum is stored etc. which must be constructed in a way that limits any risk of contamination.

Surfaces of equipment in contact with milk and colostrum are to be adequately cleaned and disinfected where necessary after use.

B. Hygiene during milking, collection and transport
1. It states that milking needs to be carried out hygienically, ensuring that before milking starts, the teats, udder and adjacent parts are cleaned. The animal is to be checked for any abnormalities and those showing clinical signs of udder

disease should not be used. Also, colostrum should be milked separately and not mixed together with raw milk.

2. Immediately after milking, milk and colostrum must be held in a clean place designed and equipped to avoid contamination. The requirements are that:

    (a) Milk must be cooled immediately to not more than 8 °C (if a daily collection), or not more than 6 °C (if collection is not daily).

    (b) Colostrum must be stored separately and immediately cooled to not more than 8 °C (if a daily collection), not more than 6 °C (if collection is not daily), or frozen.

3. During transport the cold chain must be maintained and, on arrival at the establishment of destination, the temperature of the milk and the colostrum must not be more than 10 °C.

4. Food business operators need not comply with the temperature requirements laid down in points 2 and 3 if the milk meets the criteria provided for in Part III and either:

    (a) the milk is processed within two hours of milking; or

    (b) a higher temperature is necessary for technological reasons concerning the manufacture of certain dairy products and the competent authority so authorises.

C. Staff hygiene
Those milking and/or handling raw milk and colostrum must wear suitable clean clothes. Additionally, those performing milking must maintain a high degree of personal cleanliness.

9.4.2.1.3 Criteria for raw milk

Criteria for raw milk has been made pending the establishment of standards in the context of more specific legislation on the quality of milk and dairy products. National criteria for colostrum, as regards plate count, somatic cell count or antibiotic residues, apply pending the establishment of specific Community legislation.
    A representative number of samples of raw milk and colostrum collected from milk production holdings taken by random sampling must be checked for compliance with the following in the case of raw milk and with the existing national criteria referred to for colostrums.
(i) Raw cows' milk must meet the following standards:
        Plate count 30 °C (per ml)   < or = 100,000[1]
        Somatic cell count (per ml)   < or = 400,000[2]

[1] Rolling geometric average over a two-month period, with at least two samples per month.

[2] Rolling geometric average over a three-month period, with at least one sample per month, unless the competent authority specifies another methodology to allow for seasonal variations in levels of production.

(ii)   Raw milk from other species
         Plate count 30 °C (per ml)   < or = 1,500,000[1]

[1] Rolling geometric average over a two-month period, with at least two samples per month.

However, if raw milk from species other than cows is used for manufacture of products made with raw milk by a process that doesn't involve any heat treatment, food business operators should aim to ensure the raw milk meets the following criterion:
         Plate count 30 °C (per ml)   < or = 500,000[1]

[1] Rolling geometric average over a two-month period, with at least two samples per month.

Without prejudice to Directive 96/23/EC, food business operators may not place raw milk on the market if it contains antibiotic residues in a quantity that, in respect of any one of the substances referred to in Annexes I and III to Regulation (EEC) No. 2377/90, exceeds the levels authorised under that Regulation or, if the combined total of residues of antibiotic substances exceeds any maximum permitted value.

Raw milk not complying with the above should be notified to the competent authority and action taken to correct the situation.

The checks for compliance may be carried out by, or on behalf of:

(a)   the food business operator producing the milk;

(b)   the food business operator collecting or processing the milk;

(c)   a group of food business operators; or

(d)   in the context of a national or regional control scheme.

9.4.2.2   Chapter II: Requirements Concerning Dairy and Colostrum Products

9.4.2.2.1 Temperature requirements

Food business operators are to ensure that upon acceptance at a processing establishment:

(a)   milk is quickly cooled to not more than 6 °C

(b)   colostrum is quickly cooled to not more than 6°C or maintained frozen and kept at that temperature until processed.

Note: Food business operators may keep milk and colostrum at a higher temperature if processing begins immediately after milking, within four hours of acceptance at the processing establishment, or if the competent authority authorises a higher temperature for technological reasons concerning the manufacture of certain dairy or colostrum-based products.

9.4.2.2.2 Requirements for heat treatment

1.When raw milk, colostrum, colostrum-based or dairy products undergo heat treatment, food business operators must ensure that this satisfies the requirements of Regulation (EC) No. 852/2004, Annex II, Chapter XI. In particular, when using the following processes, they should comply with the specifications mentioned:

(a) Pasteurisation is achieved by a treatment involving:

(i) a high temperature for a short time (at least 72 °C for 15 seconds);

(ii) a low temperature for a long time (at least 63 °C for 30 minutes); or

(iii) any other combination of time-temperature conditions to obtain an equivalent effect.

The result is a product that should show, where applicable, a negative reaction to an alkaline phosphatase test immediately after such treatment.

(b) Ultra high temperature (UHT) treatment is achieved by a treatment:

(i) involving a continuous flow of heat at a high temperature for a short time (not less than 135 °C in combination with a suitable holding time), such that there are no viable microorganisms or spores capable of growing in the treated product when kept in an aseptic closed container at ambient temperature; and

(ii) sufficient to ensure that the products remain microbiologically stable after incubating for 15 days at 30 °C or for 7 days at 55 °C in closed containers, or after any other method demonstrating that the appropriate heat treatment has been applied.

2. In deciding whether to subject raw milk and colostrum to heat treatment, food business operators must consider procedures developed in accordance with the HACCP principles in Regulation (EC) No. 854/2004 and comply with any requirements that the competent authority may impose in this regard when approving establishments or carrying out checks following Regulation (EC) No. 854/2004.

9.4.2.2.3 Criteria for raw cows' milk

1. Food business operators manufacturing dairy products must initiate procedures to ensure that immediately before being heat treated and if its period of acceptance specified in the HACCP-based procedures is exceeded:

(a) raw cows' milk used to prepare dairy products has a plate count of less than 300,000 per ml at 30 °C; and

(b) heat treated cows' milk used to prepare dairy products has a plate count at 30 °C of less than 100,000 per ml.

Any raw milk not complying with the above should be notified to the competent authority and action taken to correct the situation.

9.4.2.3   Chapter III: Wrapping and packaging

Consumer packages must be sealed immediately after filling in the establishment where the last heat treatment of liquid dairy products and colostrum-based products takes place using sealing devices which prevent contamination. The sealing system must be designed so that after opening, evidence of its opening remains clear and is easy to check.

9.4.2.4   Chapter IV: Labelling

1. Firstly the requirements of Directive 2000/13/EC should be met, except in the cases envisaged in Article 13(4) and (5) of that Directive. Labelling must clearly show in the case of:

(a)   raw milk for direct human consumption, the words 'raw milk';

(b)   products made with raw milk, the manufacturing process for which does not include any heat treatment or any physical or chemical treatment, the words 'made with raw milk';

(c)   colostrum, the word 'colostrum';

(d)   products made with colostrum, the words 'made with colostrum'.

2. The requirements of point 1 apply to products destined for retail trade. 'Labelling' includes any packaging, document, notice, label, ring or collar accompanying or referring to such products.

9.4.2.5   Chapter V: Identification marking

By way of derogation from the requirements of Annex II, Section I:
1. rather than indicating the approval number of the establishment, the identification mark may include a reference to where on the wrapping or packaging the approval number of the establishment is indicated;
2. in the case of the reusable bottles, the identification mark may indicate only the initials of the consigning country and the approval number of the establishment.

Note: Annex II, Section I lays down requirements for the application of the identification mark to include that it must be applied before the product leaves the establishment and if a product's packaging and/or wrapping is removed or it is further processed in another establishment, a new mark must be applied to the product. Here, the new mark must indicate the approval number of the establishment where these operations take place. The form of the identification mark is also specified to include the country code for where the establishment is located, for example, UK.

9.5 Regulation (EC) No. 854/2004 of the European Parliament and of the Council Laying Down Specific Rules for the Organisation of Official Controls on Products of Animal Origin Intended for Human Consumption

Regulation (EC) 854/2004 gives requirements for official controls on products of animal origin and states requirements for those enforcing the provisions.

In this regulation, general principles for official controls in respect of all products of animal origin falling within the scope of the regulation are given. It is a requirement that food business operators give assistance to ensure that official controls carried out by the competent authority can be done properly.

The official controls include audits of good hygiene practices and hazard analysis and critical control point (HACCP)-based procedures.

Raw milk and dairy products need to comply with the requirements of Annex IV of the regulation. This refers to the control of milk and colostrums production holdings, and to the control of raw milk and colostrums upon collection to ensure that hygiene requirements are being complied with.

9.6 Regulation (EC) No. 2073/2005 on Microbiological Criteria for Foodstuffs

Regulation (EC) No. 2073/2005 which has applied since 1 January 2006 establishes microbiological criteria for a range of foods.

The aim of this legislation is to complement food hygiene requirements, ensuring that foods being placed on the market do not pose a risk to human health, and it applies to all businesses involved in food production and handling.

The definition of 'microbiological criterion' means a criterion defining the acceptability of a product, a batch of foodstuffs or a process, based on the absence, presence or number of microorganisms, and/or on the quantity of their toxins or metabolites, per unit(s) of mass, volume, area or batch.

Two kinds of criteria have been established: *food safety criteria*, applying to products placed on the market, and *process hygiene criteria* that are applied during the manufacturing process.

9.6.1   *Food safety criteria*

Chapter 1 of the regulation focuses on food safety criteria which covers foods such as ready to eat foods intended for infants and for special medical purposes, and for milk powder and whey powder. The relevant criteria are as follows:

TABLE 9.I
Food Safety Criteria

Food Category	Microorganisms	Sampling plan[1]		Limit[2]		Analytical reference method[3]	Stage where the criterion applies
		n	c	m	M		
1.1 Ready-to-eat foods intended for infants and ready-to-eat foods for special medical purposes[4]	Listeria monocytogenes	10	0	Absence in 25 g		EN/ISO 11290-1	Products placed on the market during their shelf-life
1.2 Ready-to-eat foods able to support the growth of L. monocytogenes, other than those intended for special medical purposes	Listeria monocytogenes	5	0	100 cfu/g[5]		EN/ISO 11290-2[6]	Products placed on the market during their shelf-life
		5	0	Absence in 25 g[7]		EN/ISO 11290-1	Before the food has left the immediate control of the food business operator who has produced it
1.3 Ready-to-eat foods unable to support the growth of L. monocytogenes other than those intended for special medical purposes[4,8]	Listeria monocytogenes	5	0	100 cfu/g[5]		EN/ISO 11290-2[6]	Products placed on the market during their shelf-life
1.11 Cheeses, butter and cream made from raw milk or milk that has undergone a lower heat treatment than pasteurisation[10]	Salmonella	5	0	Absence in 25 g		EN/ISO 6579	Products placed on the market during their shelf-life
1.12 Milk powder and whey powder	Salmonella	5	0	Absence in 25 g		EN/ISO 6579	Products placed on the market during their shelf-life
1.13 Ice cream[11], excluding products where the manufacturing process or composition of the product will eliminate the salmonella risk	Salmonella	5	0	Absence in 25 g		EN/ISO 6579	Products placed on the market during their shelf-life
1.21 Cheeses, milk powder and whey powder, as referred to in the coagulase-positive staphylococci criteria in Chapter 2.2 of this Annex	Staphylococcal enterotoxins	5	0	Not detected in 25 g		European screening method of the CRL for coagulase-positive staphylococci[13]	Products placed on the market during their shelf-life
1.22 Dried infant formulae and dried dietary foods for specific medical purposes intended for infants below six months of age	Salmonella	30	0	Absence in 25 g		Absence in 25 g	Products placed on the market during their shelf-life
1.23 Dried follow-on formulae	Salmonella	30	0	Absence in 25 g		EN/ISO 6579	Products placed on the market during their shelf-life
1.24 Dried infant formulae and dried dietary foods for specific medical purposes intended for infants below six months of age[14]	Enterobacter sakazakii*	30	0	Absence in 10 g		ISO/TS 22964	Products placed on the market during their shelf-life

[1] n = number of units comprising the sample; c = number of sample units giving values between m and M

[2] For points 1.1-1.25 m = M.

[3] The most recent edition of the standard shall be used.

[4] Regular testing against the criterion is not required in normal circumstances for the following ready-to-eat foods:
- those which have received heat treatment or other processing effective to eliminate *L. monocytogenes*, when recontamination is not possible after this treatment (for example, products heat treated in their final package)
  - fresh, uncut and unprocessed vegetables and fruits, excluding sprouted seeds
  - bread, biscuits and similar products
  - bottled or packed waters, soft drinks, beer, cider, wine, spirits and similar products
  - sugar, honey and confectionery, including cocoa and chocolate products
  - live bivalve molluscs

[5] This criterion shall apply if the manufacturer is able to demonstrate, to the satisfaction of the competent authority, that the product will not exceed the limit 100 cfu/g throughout the shelf-life. The operator may fix intermediate limits during the process that must be low enough to guarantee that the limit of 100 cfu/g is not exceeded at the end of shelf-life.

[6] 1 ml of inoculum is plated on a Petri dish of 140 mm diameter or on three Petri dishes of 90 mm diameter.

[7] This criterion shall apply to products before they have left the immediate control of the producing food business operator, when he is not able to demonstrate, to the satisfaction of the competent authority, that the product will not exceed the limit of 100 cfu/g throughout the shelf-life.

[8] Products with pH $\leq$ 4.4 or $a_w \leq 0.92$, products with pH $\leq$ 5.0 and $a_w \leq 0.94$, products with a shelf-life of less than five days shall be automatically considered to belong to this category. Other categories of products can also belong to this category, subject to scientific justification.

[10] Excluding products when the manufacturer can demonstrate to the satisfaction of the competent authorities that, due to the ripening time and a_w of the product where appropriate, there is no *Salmonella* risk.

[11] Only ice creams containing milk ingredients.

[13] Reference: Community reference laboratory for coagulase-positive staphylococci. European screening method for the detection of Staphylococcal enterotoxins in milk and milk products.

[14] Parallel testing for Enterobacteriaceae and *E. sakazakii* shall be conducted, unless a correlation between these microorganisms has been established at an individual plant level. If Enterobacteriaceae are detected in any of the product samples tested in such a plant, the batch must be tested for *E. sakazakii*. It shall be the responsibility of the manufacturer to demonstrate to the satisfaction of the competent authority whether such a correlation exists between Enterobacteriaceae and *E. sakazakii*.

134

C. sakazakii is still refered to as *E. sakazakii* in legislation despite the name change in 2008.

Interpretation of the test results relating to Table 9.1

The limits given refer to each sample unit tested.

The test results demonstrate the microbiological quality of the batch tested. They may also be used for demonstrating the effectiveness of the hazard analysis and critical control point principles or good hygiene procedure of the process.

L. monocytogenes in ready-to-eat foods intended for infants and for special medical purposes:

- satisfactory, if all the values observed indicate the absence of the bacterium,

- unsatisfactory, if the presence of the bacterium is detected in any of the sample units.

L. monocytogenes in ready-to-eat foods able to support the growth of *L. monocytogenes* before the food has left the immediate control of the producing food business operator when he is not able to demonstrate that the product will not exceed the limit of 100 cfu/g throughout the shelf-life:

- satisfactory, if all the values observed indicate the absence of the bacterium,

- unsatisfactory, if the presence of the bacterium is detected in any of the sample units.

L. monocytogenes in other ready-to-eat foods:

- satisfactory, if all the values observed are ≤ the limit,

- unsatisfactory, if any of the values are > the limit.

Salmonella in different food categories:

- satisfactory, if all the values observed indicate the absence of the bacterium,

- unsatisfactory, if the presence of the bacterium is detected in any of the sample units.

Staphylococcal enterotoxins in dairy products:

- satisfactory, if in all the sample units the enterotoxins are not detected,

- unsatisfactory, if the enterotoxins are detected in any of the sample units.

Enterobacter sakazakii in dried infant formulae and dried dietary foods for special medical purposes intended for infants below 6 months of age:

- satisfactory, if all the values observed indicate the absence of the bacterium,

- unsatisfactory, if the presence of the bacterium is detected in any of the sample units.

9.6.2   Process hygiene criteria

Chapter 2 focuses on process hygiene criteria, with chapter 2.2 referring to milk and dairy products. Food categories covered range from pasteurised milk to ice cream and frozen dairy desserts. The criteria are outlined in Table 9.II overleaf.

Interpretation of the test results relating to Table 9.II

The limits given refer to each sample unit tested.
   The test results demonstrate the microbiological quality of the process tested.

Enterobacteriaceae in dried infant formulae, dried dietary foods for special medical purposes intended for infants below six months of age and dried follow-on formulae:

   - satisfactory, if all the values observed indicate the absence of the bacterium,

   - unsatisfactory, if the presence of the bacterium is detected in any of the sample units.

E. coli, Enterobacteriaceae (other food categories) and coagulase-positive staphylococci:

   - satisfactory, if all the values observed are $\leq$ m,

   - acceptable, if a maximum of c/n values are between m and M, and the rest of the values observed are $\leq$ m,

   - unsatisfactory, if one or more of the values observed are > M or more than c/n values are between m and M.

Presumptive *Bacillus cereus* in dried infant formulae and dried dietary foods for special medical purposes intended for infants below six months of age:

   - satisfactory, if all the values observed are $\leq$ m,

   - acceptable, if a maximum of c/n values are between m and M, and the rest of the values observed are $\leq$ m,

   - unsatisfactory, if one or more of the values observed are > M or more than c/n values are between m and M.

TABLE 9.II
Process Hygiene Criteria

Food Category	Microorganisms	Sampling plan[1]		Limit[2]		Analytical reference method[3]	Stage where the criterion applies	Action in case of unsatisfactory results
		n	c	m	M			
2.2.1 Pasteurised milk and other pasteurised liquid dairy products[4]	Entero-bacteriaceae	5	2	<1/ml	5/ml	ISO 21528-1	End of the manufacturing process	Check on the efficiency of heat treatment and prevention of recontamination, as well as the quality of raw materials
2.2.2 Cheeses made from milk or whey that has undergone heat treatment	E-coli[5]	5	2	100 cfu/g	1000 cfu/g	ISO 16649-1 or 2	At the time during the manufacturing process when the E. coli count is expected to be highest[6]	Improvements in production hygiene and selection of raw materials
2.2.3 Cheeses made from raw milk	Coagulase-positive staphylococci	5	2	10^4 cfu/g	10^5 cfu/g	EN/ISO 6888-2	At the time during the manufacturing process when the number of staphylococci is expected to be highest	Improvements in production hygiene and selection of raw materials. If values >10^5 cfu/g are detected, the cheese batch has to be tested for staphylococcal enterotoxins
2.2.4 Cheeses made from raw milk that has undergone a lower heat treatment than pasteurisation[7] and ripened cheeses made from milk or whey that has undergone pasteurisation or a stonger heat treatment	Coagulase-positive staphylococci	5	2	100 cfu/g	1000 cfu/g	EN/ISO 6888-1 or 2		

137

TABLE 9.II cont.
Process Hygiene Criteria

Food Category	Microorganisms	Sampling plan[1]		Limit[2]		Analytical reference method[3]	Stage where the criterion applies	Action in case of unsatisfactory results
		n	c	m	M			
2.2.5 Unripened soft cheeses (fresh cheeses) made from milk or whey that has undergone pasteurisation or a stronger heat treatment[7]	Coagulase-positive staphylococci	5	2	10 cfu/g	100 cfu/g	EN/ISO 6888-1 or 2	End of the manufacturing process	Improvements in production hygiene. If values $>10^5$ cfu/g are detected, the cheese batch has to be tested for staphylococcal enterotoxins
2.2.6 Butter and cream made from raw milk or milk that has undergone a lower heat treatment than pasteurisation	E. coli[5]	5	2	10 cfu/g	100 cfu/g	ISO 16649-1 or 2	End of the manufacturing process	Improvements in production hygiene and selection of raw materials
2.2.7 Milk powder and whey powder[4]	Enterobacteriaceae	5	0	10 cfu/g		ISO 21528-2	End of the manufacturing process	Check on efficiency of heat treatment and prevention of recontamination
	Coagulase-positive staphylococci	5	2	10 cfu/g	100 cfu/g	EN/ISO 6888-1 or 2	End of the manufacturing process	Improvements in production hygiene. If values $>10^5$ cfu/g are detected, the batch has to be tested for staphylococcal enterotoxins
2.2.8 Ice cream and[8] frozen dairy desserts	Enterobacteriaceae	5	2	10 cfu/g	100 cfu/g	ISO 21528-2	End of the manufacturing process	Improvements in production hygiene

TABLE 9.II cont.
Process Hygiene Criteria

Food Category	Microorganisms	Sampling plan[1]		Limit[2]		Analytical reference method[3]	Stage where the criterion applies	Action in case of unsatisfactory results
		n	c	m	M			
fomulae and dried dietary foods for special medical purposes intended for infants below six months of age	Entero-bacteriaceae	10	0	Absence in 10 g		ISO 21528-1	End of the manufacturing process	Improvements in production hygiene to minimise contamination[9]
2.2.10 Dried follow-on fomulae	Entero-bacteriaceae	5	0	Absence in 10 g		ISO 21528-1	End of the manufacturing process	Improvements in production hygiene to minimise contamination
2.2.11 Dried infant fomulae and dried dietary foods for special medical purposes intended for infants below six months of age	Presumptive *Bacillus cereus*	5	1	50 cfu/g	500 cfu/g	EN/ISO 7932[10]	End of the manufacturing process	Improvements in production hygiene. Prevention of recontamination. Selection of raw material

139

[1] n = number of units comprising the sample; c = number of sample units giving values between m and M

[2] For these criteria 2.2.7, 2.2.9 and 2.2.10 m=M

[3] The most recent edition of the standard shall be used.

[4] The criterion does not apply to milk destined for further processing in food industry.

[5] *E. coli* is used here as an indicator for the level of hygiene.

[6] For cheeses which are not able to support the growth of *E. coli*, the *E. coli* count is usually the highest at the beginning of the ripening period, and for cheeses which are able to support the growth of *E. coli*, it is normally at the end of the ripening period.

[7] Excluding cheeses where the manufacturer can demonstrate, to the satisfaction of the competent authorities, that the product does not pose a risk of staphylococcal enterotoxins.

[8] Only ice creams containing milk ingredients.

[9] Parallel testing for Enterobacteriaceae and *E. sakazakii* shall be conducted, unless a correlation between these micro-organisms has been established at an individual plant level. If Enterobacteriaceae are detected in any of the product samples tested in such a plant, the batch has to be tested for *E. sakazakii*. It shall be the responsibility of the manufacturer to demonstrate to the satisfaction of the competent authority whether such a correlation exists between Enterobacteriaceae and *E. sakazakii*.

[10] 1 ml of inoculum is plated on a Petri dish of 140 mm diameter or on three Petri dishes of 90 mm diameter.

9.7 Food Hygiene (England) Regulations 2006, S.I. 2006 No. 14 (Hygiene requirements specific to the UK)

9.7.1 Sale of raw milk intended for direct human consumption

Regulation 32 of the Food Hygiene (England) Regulations 2006, S.I. 2006 No. 14, requires that Schedule 6 concerning restrictions on the sale of raw milk intended for direct human consumption shall have effect. The provisions of this Schedule are as follows:

1. It is an offence to sell raw milk intended for direct human consumption if it does not comply with the following standards:

Plate count at 30 °C (cfu per ml)	< or = 20,000
Coliforms (cfu per ml)	< 100

2. Only the occupier of a production holding or a distributor in compliance with the stated requirements may sell raw cows' milk intended direct for human consumption.

3. The occupier of a production holding may only sell raw cows' milk intended direct for human consumption:

a) at or from the farm premises where the animals from which the milk has been derived are maintained; and

b) to:

(i) the final consumer for consumption other than at those farm premises;

(ii) a temporary guest or visitor to those farm premises as or part of a meal or refreshment; or

(iii) a distributor.

4. A distributor may only sell raw cows' milk intended direct for human consumption:

a) which he has bought as in point 3 above;

b) in the containers in which he receives the milk, with the container fastenings unbroken;

c) from a vehicle lawfully used as a shop premises;

d) direct to the final consumer.

5. Where the farm premises are being used for the sale of raw cows' milk intended for direct human consumption, the Food Standards Agency shall carry out such sampling, analysis and examination of the milk as it considers necessary to ensure it meets the required standards. A stated fee applies.

9.7.2 *Temperature control requirements*

In the UK, Schedule 4 of the Food Hygiene (England) Regulations 2006, S.I. 2006 No. 14 details temperature control requirements for foods in general.

The regulations prescribe a chilled food holding temperature of 8 °C or less, but there is also a general requirement that foods must not be kept at temperatures that would result in a risk to health, and particularly that perishable foodstuffs must not be kept at above the maximum recommended storage temperature, which overrides the 8 °C requirement. Hot-held foods (food having been cooked or reheated that is for service or on display for sale) must not be kept below 63 °C.

The regulations provide for defences in relation to upward variations of the 8 °C temperature, tolerance periods for chill-holding of foods and hot-holding variations. The defendant may be required to produce well-founded scientific proof to support his claims. For example, with chill-holding tolerance periods, the defendant will need to prove that the food was on service or display, had not been previously put on display at more than 8 °C and had been kept there for less than four hours. Alternatively, it would need to be proved that the food was being transferred to or from a vehicle used for the activities of a food business, to or from premises (including vehicles) at which the food was to be kept at or below 8°C or the recommended temperature, or, was kept at above 8°C or the recommended temperature for an unavoidable reason, such as that below, and was

kept at above 8 °C or the recommended temperature for a limited period consistent with food safety. The permitted reasons are given below:
- to facilitate handling during and after processing or preparation
- the defrosting of equipment, or
- temporary breakdown of equipment

For Scotland, there are separate provisions to include requirements to hold food under refrigeration or in a cool ventilated place, or at a temperature above 63 °C and to reheat food to a temperature of at least 82 °C (The Food Hygiene (Scotland) Regulations 2006, S.S.I. 2006 No. 3).

Schedule 4 of the Food Hygiene (England) Regulations 2006 contains several definitions, including:

Shelf-life: where the minimum durability or 'use by' indication is required according to Regulation 20 or 21 of the Food Labelling Regulations 1996 (form of indication of minimum durability and form of indication of 'use-by date'), the period up to and including that date. For other food, the period for which it can be expected to remain fit for sale when kept in a manner consistent with food safety.

Recommended temperature: a specified temperature that has been recommended in accordance with a food business responsible for manufacturing, preparing or processing the food recommending that it be kept at or below a specified temperature between 8 °C and ambient temperatures.

It should be noted that the temperature control requirements as detailed in Schedule 4 of the Food Hygiene (England) Regulations 2006 (S.I. 2006 No. 14) do not apply to any food covered by EU Regulation 853/2004 on hygiene of products of animal origin or any food business operation carried out on a ship or aircraft.

9.8   Guidance

In the U.K, the Food Standards Agency has published guidance notes on the requirements of the EU hygiene and microbiological criteria regulations which can, at the time of going to press, be accessed at the following link: http://www.food.gov.uk/foodindustry/guidancenotes/hygguid/fhlguidance/

9.9   References

1. Regulation (EC) No. 852/2004 of the European Parliament and of the Council on the hygiene of foodstuffs (OJ No. L139, 30.4.2004, 1). The revised text of Regulation (EC) No. 852/2004 is now set out in a Corrigendum (OJ No. L226, 25.6.2004, 3) as amended by Regulation 1019/2008 and as read with Regulation 2073/2005.

2. Regulation (EC) No. 853/2004 of the European Parliament and of the Council laying down specific hygiene rules for food of animal origin (OJ No. L139, 30.4.2004, p.55). The revised text of Regulation (EC) No. 853/2004 is now set out in a Corrigendum (OJ No. L226, 25.6.2004, p.22) as amended by Regulation 2074/2005, Regulation 2076/2005, Regulation 1662/2006, Regulation 1791/2006 and Regulation 1020/2008 and as read with Directive 2004/41, Regulation 1688/2005, Regulation 2074/2005 and Regulation 2076/2005.

3. Regulation (EC) No. 854/2004 of the European Parliament and of the Council laying down specific rules for the organisation of official controls on products of animal origin intended for human consumption(OJ No. L139, 30.4.2004, p.206). The revised text of Regulation (EC) No. 854/2004 is now set out in a Corrigendum (OJ No. L226, 25.6.2004, p.83) as amended by Regulation 882/2004, Regulation 2074/2005, Regulation 2076/2005, Regulation 1663/2006, Regulation 1791/2006 and Regulation 1021/2008 and as read with Directive 2004/41, Regulation 2074/2005, Regulation 2075/2005 and Regulation 2076/2005.

4. Commission Regulation (EC) No. 2073/2005 on microbiological criteria for foodstuffs (OJ No. L338, 22.12.2005, p.1, as read with the corrigenda at OJ No. L283, 14.10.2006, p.62) as amended by Regulation 1441/2007.

10. PATHOGEN PROFILES

10.1   *Bacillus cereus*

10.1.1   *Morphology*

Gram-positive spore-forming rods; 1.0 - 1.2 x 3.0 - 7.0 μm.

10.1.2   *Oxygen requirements*

Facultative anaerobe - normally aerobic.

10.1.3   *Temperature*

Typically, the vegetative cells of *B. cereus* have an optimum growth temperature of 30 - 35 °C, and a maximum ranging from 48 - 55 °C (1, 2, 3). However, psychrotrophic strains have been identified - especially in milk and dairy products - capable of growing within the range 4 - 37 °C (4). Most of these strains were also reported as capable of producing enterotoxin at 4 °C after prolonged incubation (>21 days) (3, 5).

10.1.4   *Heat resistance*

Vegetative cells of *B. cereus* are readily destroyed by pasteurisation or equivalent heat treatments. However, spores can survive quite severe heat processes, but there is considerable variation between different strains. D_{95} - values of between 1.2 and 36 minutes have been reported (6). It has been shown that strains commonly implicated in food poisoning are more heat-resistant than other strains, and are therefore more likely to survive a thermal process.

10.1.5   *pH*

B. cereus has been reported to be capable of growth at pH values between 4.3 and 9.3, under otherwise ideal conditions (6, 7).

10.1.6 A_w

The minimum water activity in which *B. cereus* has been reported to grow is 0.95; possibly as low as 0.91 (in fried rice) or less (6).

10.1.7   *Characteristics of B. cereus toxins*

The emetic toxin of *B. cereus* is stable in the pH range 2 - 11 (1); it is also heat-resistant, and able to resist heating to 126 °C for 90 min (3, 6). This toxin is produced after active (vegetative) growth at the end of the growth cycle, and may be associated with the formation of spores.

The diarrhoeal enterotoxin is unstable at pH values of < 4.0 or > 11.0 (6), and is heat-sensitive, being destroyed at 56 °C for 5 minutes (1, 6). The toxin is a protein that is produced during active growth.

10.2   *Campylobacter* spp.

10.2.1   *Morphology*

Gram-negative spirally curved rods; 0.2 - 0.8 x 0.5 - 5.0 μm.

10.2.2   *Oxygen requirements*

Campylobacter is both microaerophilic and 'capnophilic' (liking carbon dioxide); its growth is favoured by an atmosphere containing 10% carbon dioxide and 5 - 6% oxygen. Growth is also enhanced by hydrogen. The organism will normally die rapidly in the presence of air; it is particularly sensitive to oxygen breakdown products. Because of this and other growth characteristics (see below), these organisms are not normally capable of growing in foods.

10.2.3   *Temperature*

Campylobacter jejuni and *Campylobacter coli* only grow at temperatures above about 30 °C; they (and *Campylobacter lari*) are consequently referred to as the thermophilic group of *Campylobacters*. Their optimum temperature for growth is between 42 and 43 °C, with a maximum of 45 °C (8).

Campylobacter survives poorly at room temperatures (around 20 - 23 °C); it dies much more quickly than at refrigeration temperatures. It can survive well for short periods at chill temperatures. On the other hand, it is generally more sensitive to freezing, although there may be some survival for long periods (9, 10, 11).

10.2.4   Heat resistance

C. jejuni is very heat-sensitive. Heat injury can occur at 46 °C or higher. z-values range from 48 - 60 °C depending on pH (8). D-values of 7.2 - 12.8 min have been reported at 48 °C (in skimmed milk) (8). At 55 °C, the range was 0.74 - 1.0 (8). The organism cannot survive normal milk pasteurisation.

10.2.5   pH

Campylobacter has an optimum pH for growth in the range 6.5 - 7.5 and no growth is observed below pH 4.9 (8).

10.2.6 A_w/Sodium chloride

Campylobacter is particularly sensitive to drying; it does not survive well in dry environments. The minimum water activity for growth is 0.98. *Campylobacter* is also quite sensitive to sodium chloride (NaCl); levels of 2% or more can be bactericidal to the organism. The effect is temperature-dependent; the presence of even 1% NaCI can be inhibitory or bactericidal, depending on temperature. The bactericidal effect decreases with decreasing temperature (12).

10.3   Clostridium botulinum

Seven different types of *C. botulinum* are known, forming at least seven different toxins; A to G. Types A, B, E and, to a lesser extent, F are the types that are responsible for most cases of human botulism (13, 14). All type A strains are proteolytic, and type E strains are usually non-proteolytic; types B and F can be either. There are four main groupings of the organism, and Groups I and II are those responsible for cases of botulism.

10.3.1   Morphology

Gram-positive spore-forming rods; 0.5 - 2.4 x 1.7 - 22.0 μm.

10.3.2   Oxygen requirements

Although *C. botulinum* is a strict anaerobe, many foods that are not obviously 'anaerobic' can provide adequate conditions for growth. Thus, an aerobically packed product may not support the growth of the organism on the surface, but the interior of the food may do so. It is also important to note that the inclusion of oxygen as a packaging gas cannot ensure that growth of *C. botulinum* is prevented.

10.3.3 Temperature

All strains of *C. botulinum* grow reasonably well in the temperature range of 20 - 45 °C, but the low temperatures required to inhibit Groups I (proteolytic group) and II (non-proteolytic group) are different. Group I will not grow at temperatures of 10 °C or less, but Group II strains are psychrotrophic, being capable of slow growth and toxin production at low temperatures - even as low as 3 °C (15, 16).

10.3.4 Heat resistance

The vegetative cells of *C. botulinum* are not particularly heat-resistant, but the spores of this organism are more so. All *C. botulinum* types produce heat labile toxins, which may be inactivated by heating at 80 °C for 20 - 30 min, at 85 °C for 5 min, or at 90 °C for a few seconds.

10.3.5 pH

The minimum pH for the growth of proteolytic and non-proteolytic strains is pH 4.6 and 5.0, respectively (17, 18).

10.3.6 A_w/Sodium chloride

The minimum a_w for growth of *C. botulinum* depends on solute, pH and temperature, but under optimum growth conditions 10% (w/w) NaCl is required to prevent growth of Group I, and 5% (w/w) NaCl is necessary to prevent growth of Group II organisms. These concentrations correspond to limiting a_w of 0.94 for Group I and 0.97 for Group II (13). These values have been established under carefully controlled laboratory conditions. In commercial situations, safety margins must be introduced to allow for process variability.

10.3.7 Characteristics of C. botulinum spores

The most heat-resistant spores of Group I *C. botulinum* are produced by type A and proteolytic B strains for which D values are 0.1 - 0.21 minutes at 121 °C (18).

The spores of Group II (non-proteolytic/psychotrophic) strains are less heat-resistant than Group I strains. However, they may survive mild heat treatments (70 - 85 °C) and their ability to grow at refrigeration temperatures necessitates their control in foods capable of supporting their growth (e.g. vacuum-packed, par-cooked meals with pH value > 5.0 and a_w > 0.97) (19, 20). D-values at 100 °C are < 0.1 minutes (18).

10.4   *Clostridium perfringens*

10.4.1   *Morphology*

Gram-positive spore-forming rods; 0.3 - 1.9 x 2.0 - 10.0 μm.

10.4.2   *Oxygen requirements*

C. perfringens - like other clostridia - is an anaerobe. It will not, therefore, grow on the surface of foods unless they are vacuum- or gas-packed. The organism will grow well in the centre of meat or poultry dishes, where oxygen levels are reduced, particularly by cooking.

10.4.3   *Temperature*

The most significant characteristic of *C. perfringens* in relation to food safety is the organism's ability to grow extremely rapidly at high temperatures. Its optimum temperature range for growth is 43 - 45 °C, although *C. perfringens* has the potential ability to grow within the temperature range 15 - 50 °C, depending on strain and other conditions. While some growth can occur at 50 °C, death of the vegetative cells of this organism usually occurs rapidly above this temperature (21, 22). At cold temperatures (0 - 10 °C) vegetative cells die rapidly (21).

10.4.4   *Heat resistance*

Exposure to a temperature of 60 °C or more will result in the death of vegetative cells of *C. perfringens*, although prior growth at high temperatures, or the presence of fat in a food will result in increased heat resistance. (It is unusual for spores to be formed in foods after the growth of this organism) (23). In addition, the enterotoxin is not heat-resistant; it is destroyed by heating at 60 °C for 10 minutes (23, 24, 25).

10.4.5   *pH*

C. perfringens is not a tolerant organism with respect to pH. It grows best at pH values between 6 and 7 (the same pH as most meats). Under otherwise ideal conditions, very limited growth may occur at pH values over the range pH ≤ 5 - $\geq$ 8.3. Spores, however, will survive greater extremes of pH (and a_w) (21, 22).

10.4.6   *A_w/Sodium chloride*

C. perfringens is not tolerant of low water activities. As in the case of other factors limiting the growth or survival of this organism, the limits for water activity are affected by temperature, pH, type of solute, etc. The lowest a_w recorded to support the growth of *C. perfringens* appears to be 0.93 to 0.97 depending on the solute (glycerol and sucrose respectively) used to control a_w (22, 26). Salt concentrations of 6 - 8% inhibit growth of most *C. perfringens* strains. Some studies indicate that the presence of 3% NaCl delays growth of *C. perfringens* in vacuum-packed beef (26).

10.4.7   Characteristics of C. perfringens spores

The spores of *C. perfringens* can vary quite considerably in their heat resistance, which is affected by the heating substrate. Recorded heat resistance values (D-values) at 95 °C range from 17.6 - 64.0 minutes for heat-resistant spores, to 1.3 - 2.8 minutes for heat-sensitive spores (21).

10.5   Cronobacter (Enterobacter) sakazakii

C. sakazakii is a new genus in the family Enterobacteriaceae. It is a taxanomic reclassification of the pathogen *Enterobacter sakazakii* and consists of five species; *Cronobacter sakazakii* (and includes *Cronobacter sakazakii* subsp. *sakazakii* and *Cronobacter sakazakii* subsp. *malonaticus), Cronobacter turicensis, Cronobacter muytjensii, Cronobacter dublinensis and Cronobacter genomospecies 1*. It accomodates the biogroups of *E. sakazakii* (27, 28).

10.5.1   Morphology

Gram-negative rod.

10.5.2   Oxygen requirements

C. sakazakii is a facultative anaerobe.

10.5.3   Temperature

The minimum growth temperature is between 5.5 and 8 °C. The lowest recorded temperature at which *C. sakazakii* is known to grow is 3.4 °C, suggesting that the organism is able to grow during refrigeration. The maximum growth temperature ranges in general from 41 – 45 °C.

10.5.4   Heat resistance

C. sakazakii is considered to be one of the most thermo-tolerant among the Enterobacteriaceae, as *C. sakazakii* can survive at elevated temperatures (45 °C), and has the ability to grow at temperatures up to 47 °C in warm and dry environments such as in the vicinity of drying equipment in factories. It has a competitive advantage when compared to other members of the Enterobacteriaceae. However, it does not survive a standard pasteurisation process (> 60 °C) (29, 30, 31, 32).

10.5.5 pH

Like other members of the Enterobacteriaceae, *C. sakazakii* is presumed to have good resistance to low pH. Survival of the organism in acid environments depends on a number of factors such as pH, acidulant identity, acidulant concentrations, temperature, water activity, atmosphere, and the presence of other inhibitory compounds (33).

10.5.6 A_w

C. sakazakii can survive in dried infant formula having a water activity of approximately 0.2.

10.6   *Escherichia coli* O157

10.6.1   *Morphology*

Gram-negative short rods; 1.1 - 1.5 x 2.0 - 6.0 μm.

10.6.2   *Oxygen requirements*

E. coli O157 is a facultative anaerobe; it grows well under aerobic or anaerobic conditions. High levels of carbon dioxide may inhibit its growth.

10.6.3   *Temperature*

The growth range for *E. coli* O157 is thought to be between 7 and 45°C, with an optimum of approximately 37 °C (34). (Note: *E. coli* O157:H7 grows poorly at 44 - 45 °C and does not grow within 48 hours at 45.5 °C. Therefore, traditional detection methods for *E. coli* in foods cannot be relied upon to detect *E. coli* O157:H7).

10.6.4   *Heat resistance*

E. coli O157 is not a heat-resistant organism. D-values at 57 and 63°C in meat have been reported as approximately 5 and 0.5 minutes, respectively (35).

Anaerobic growth, reduced a_w, high fat content and exposure to prior heat shock may result in higher D-values. However, it is unlikely to survive conventional milk pasteurisation.

10.6.5 pH

The minimum pH for growth, under optimal conditions, is 4.0 - 4.4 (using hydrochloric acid as an acidulant) (36, 37). The minimum value is affected by the acidulant used, with both lactic and acetic acids being more inhibitory than hydrochloric acid (34). *E. coli* O157 is unusually acid-tolerant and survives well in foods with low pH values (3.6 - 4.0), especially at chill temperatures (38).

10.6.6 A_w/ Sodium chloride

Current published data suggest that *E. coli* O157 grows well at NaCl concentrations up to 2.5% and may grow at concentrations of at least 6.5% (w/v) (a_w less than 0.97) under otherwise optimal conditions (39). The organism appears to be able to tolerate certain drying processes (38).

10.7 *Listeria* spp.

10.7.1 *Morphology*

Gram-positive short rods; 0.4 - 0.5 x 0.5 - 2.0 μm.

10.7.2 *Oxygen requirements*

Aerobe or microaerophilic.

10.7.3 *Temperature*

Listeria monocytogenes is unusual amongst foodborne pathogens in that it is psychrotrophic, being potentially capable of growing - albeit slowly - at refrigeration temperatures down to, or even below 0 °C. However, -0.4 °C is probably the most likely minimum in foods (40). Its optimum growth temperature, however, is between 30 and 37 °C; growth at low temperatures can be very slow, requiring days or weeks to reach maximum numbers. The upper temperature limit for the growth of *L. monocytogenes* is reported to be 45 °C (41).

10.7.4   Heat resistance

L. monocytogenes is not a particularly heat-resistant organism; it is not a spore-former, so can be destroyed by pasteurisation. It has been reported to have slightly greater heat resistance than certain other foodborne pathogens. It is generally agreed that milk pasteurisation will destroy normal levels of *L. monocytogenes* in milk ($>10^5$/ml); the D-value is 0.1 - 0.3 minutes at 70 °C in milk.

D-values at 68.9 °C for the strain Scott A were 6 seconds in raw 38% milk fat cream, and 7.8 seconds in inoculated sterile cream. z-values were 6.8 and 7.1 °C, respectively (42, 43).

10.7.5   pH

The ability of *Listeria* to grow at different pH values (as with other bacteria) is markedly affected by the type of acid used and temperature. Under ideal conditions, the organism is able to grow at pH values well below pH 5 (pH 4.3 is the lowest value where growth has been recorded, using hydrochloric acid as acidulant). In foods, however, the lowest limit for growth is likely to be considerably higher - especially at refrigeration temperatures, and where acetic acid is used as acidulant; pH < 5.2 has been suggested as the lowest working limit (44).

10.7.6 A_w/Sodium chloride

L. monocytogenes is quite tolerant of high NaCl/low a_w. It is likely to survive, or even grow, at salt levels found in foods (10 - 12% NaCl or more). It grows best at a_w of ≥ 0.97, but has been shown to be able to grow at a_w level of 0.90. The bacterium may survive for long periods at a_w as low as 0.83 (41).

10.8   Salmonella spp.

10.8.1   Morphology

Gram-negative short rods with peritrichous flagella; 0.5 - 0.7 x 1.0 - 3.0 μm.

10.8.2   Oxygen requirements

Facultative anaerobe.

10.8.3   Temperature

Salmonellae can grow in the temperature range of 7 - 48 °C. However, some strains are able to grow at temperatures as low as 4 °C (45). Growth is slow at temperatures below about 10 °C, the optimum being 35 - 37 °C.

Salmonellae are quite resistant to freezing, *Salmonella enteritidis* were isolated from ice cream held at -23 °C for 7 years (46), and may survive in some foods for a number of years.

10.8.4   Heat resistance

Salmonella is not a spore-forming organism. It is not, therefore, a heat-resistant organism; pasteurisation and equivalent treatments will destroy the organism under normal circumstances. D values normally range from about 1 to 10 min at 60 °C, with a z-value of 4 - 5 °C. However, high fat or low a_w will reduce the effectiveness of heat treatments, and appropriate heat treatments must be determined experimentally for low a_w foods. Furthermore, strains vary in their ability to withstand heating; *Salmonella senftenberg* 775W is about 10 to 20 times more heat-resistant than the average strain of *Salmonella* at high a_w (47). The D-value for *S. senftenberg* in milk at 71.7 °C is 0.02 minutes, and the D-value for *Salmonella* spp. in milk at 68.3 °C is 0.01 minutes (47).

10.8.5   pH

Salmonella has a pH range for growth of 3.8 - 9.5, under otherwise ideal conditions, and with an appropriate acid. Some death will occur at pH values of less than about 4.0, depending on the type of acid and temperature. The optimal pH for *Salmonella* growth is between 6.5 - 7.5.

10.8.6 A_w/Sodium chloride

Where all other conditions are favourable, *Salmonella* has the potential to grow at a_w levels as low as 0.945, or possibly 0.93 (as reported in dried meat and dehydrated soup), depending on serotype, substrate, temperature and pH. Salmonellae are quite resistant to drying.

The growth of *Salmonella* is generally inhibited by the presence of 3 - 4% NaCl, although salt tolerance increases with increasing temperature (48).

10.9   *Staphylococcus aureus*

10.9.1   Morphology

Gram-positive cocci; 0.7 - 0.9 μm diameter.

10.9.2 Oxygen requirements

Facultative anaerobe. The growth of *Staph. aureus* is more limited under anaerobic than under aerobic conditions. The limits for toxin production are also narrower than for growth. The following relate to limits for growth only.

10.9.3 Temperature

Under otherwise ideal conditions *Staph. aureus* can grow within the temperature range 7 - 48.5 °C, with an optimum of 30 - 37 °C (49). It can survive well at low temperatures.

Freezing and thawing have little effect on *Staph. aureus* viability, but may cause some cell damage (50).

10.9.4 Heat resistance

Heat resistance depends very much on the food type in which the organism is being heated (conditions relating to pH, fat content, water activity, etc.). As is the case with other bacteria, stressed cells can also be less tolerant of heating.

Under most circumstances, however, the organism is heat-sensitive and will be destroyed by pasteurisation. In milk, the D-value at 60 °C is 1 - 6 minutes, with a z-value of 7 - 9 °C.

10.9.5 pH

The pH at which a staphylococcal strain will grow is dependent on the type of acid (acetic acid is more effective at destroying *Staph. aureus* than citric acid), water activity and temperature (sensitivity to acid increases with temperature).

Most strains of staphylococci can grow within the pH range 4.2 to 9.3 (optimum 7.0 - 7.5), under otherwise ideal conditions (49, 51).

10.9.6 A_w/Sodium chloride

Staph. aureus is unusual amongst food-poisoning organisms in its ability to tolerate low water activities. It can grow over the a_w range 0.83 - > 0.99 aerobically under otherwise optimal conditions. However, an a_w of 0.86 is the generally recognised minimum in foods (52).

Staphylococci are more resistant to salt present in foods than other organisms. In general, *Staph. aureus* can grow in 7 - 10% salt, but certain strains can grow in 20%. An effect of increasing salt concentration is to raise the minimum pH for growth.

10.9.7 Limits permitting toxin production

Temperature: 10 - 45 °C (optimum between 35 and 40 °C) (very little toxin is produced at the upper and lower extremes) (51)

pH: 5.2 - 9.0 (optimum 7.0 and 7.5) (49, 51)

*A_w: between 0.87 and > 0.99

Atmosphere: little or no toxin production in anaerobically packed foods, especially vacuum-packed foods (53)

Heat Resistance: enterotoxins are quite heat-resistant. In general, heating at 100 °C for at least 30 minutes may be required to destroy unpurified toxin (51, 54).

* dependent on temperature, pH, atmosphere, strain, and solute.

10.10 Yersinia enterocolitica

10.10.1 Morphology

Gram-negative short rods (occasionally coccoid); 0.5 - 1.0 x 1.0 - 2.0 μm.

10.10.2 Oxygen requirements

Facultative anaerobe. Carbon dioxide has some inhibitory effect on the growth of *Y. enterocolitica*. Vacuum packaging can retard growth to a lesser extent.

10.10.3 Temperature

Yersinias are psychrotrophic organisms, being capable of growth at refrigeration temperatures. Extremely slow growth has been recorded at temperatures as low as 0 to -1.3 °C. However, the optimum temperature for growth of *Y. enterocolitica* is 28 - 29 °C with the reported growth range of -2 - 42 °C (55, 56, 57). The maximum temperature where growth has been recorded is 44 °C (57, 58).

The organism is quite resistant to freezing and has been reported to survive in frozen foods for long periods (55, 56).

10.10.4 Heat resistance

The organism is sensitive to heat, being easily killed at temperatures above about 60 °C. D-values of between 0.18 and 0.96 minutes at 62.8 °C in milk have been reported (57); D-values in scaling water were 96, 27 and 11 seconds at 58 °C, 60 °C and 62 °C respectively (56). It will therefore be destroyed by standard milk pasteurisation (55).

10.10.5 pH

Yersinia is sensitive to pH values of less than 4.6 (more typically 5.0) in the presence of organic acids, e.g. acetic acid. *Y. enterocolitica* are not able to grow at pH < 4.2 or > 9.0. A lower pH minimum for growth (pH 4.1 - 4.4) has been observed with inorganic acids, under otherwise optimal conditions. Its optimum is pH 7.0 - 8.0; they tolerate alkaline conditions extremely well (59).

10.10.6 A_w/Sodium chloride

Yersinia may grow at salt concentrations up to about 5% (a_w 0.96), but no growth occurs at 7% (a_w 0.945). Growth is retarded in foods containing 5% salt (57, 59).

10.11   References

1. Rajkowski K.T., Bennett R.W. *Bacillus cereus*, in *International Handbook of Foodborne Pathogens*. Eds. Miliotis M.D., Bier J.W. New York, Marcel Dekker. 2003, 27-40.

2. Fermanian C., Fremy J.M., Claisse M. Effect of temperature on the vegetative growth of type and field strains of *Bacillus cereus*. *Letters in Applied Microbiology*, 1994, 19 (6), 414-8.

3. International Commission on Microbiological Specifications for Foods. *Bacillus cereus*, in *Microorganisms in Foods, Volume 5: Microbiological Specifications of Food Pathogens*. Ed. International Commission on Microbiological Specifications for Foods. London, Blackie. 1996, 20-35.

4. van Netten P., van de Moosdijk A., van Hoensel P., Mossel D.A.A., Perales I. Psychrotrophic strains of *Bacillus cereus* producing enterotoxin. *Journal of Applied Bacteriology*, 1990, 69 (1), 73-9.

5. Dufrenne J., Soentoro P., Tatini S., Day T., Notermans S. Characteristics of *Bacillus cereus* related to safe food production. *International Journal of Food Microbiology*, 1994, 14 (2), 87.

6. Kramer J.M., Gilbert R.J. *Bacillus cereus* and other *Bacillus* species, in *Foodborne Bacterial Pathogens*. Ed. Doyle M.P. New York, Marcel Dekker. 1989, 21-70.

7. Fermanian C., Fremy J.-M., Lahellec C. *Bacillus cereus* pathogenicity: a review. *Journal of Rapid Methods and Automation in Microbiology*, 1993, 2 (2), 83-134.

8. International Commission on Microbiological Specifications for Foods. *Campylobacter*, in *Microorganisms in Foods, Volume 5: Microbiological Specifications of Food Pathogens*. Ed. International Commission on Microbiological Specifications for Foods. London, Blackie. 1996, 45-65.

9. Hu L. Kopecko D. J. *Campylobacter* Species, in *International Handbook of Foodborne Pathogens*. Eds. Miliotis M. D., Bier J. W. New York, Marcel Dekker. 2003, 181-98.

10. Park S. *Campylobacter*: stress response and resistance, in *Understanding Pathogen Behaviour: Virulence, Stress Response and Resistance*. Ed. Griffiths M. Cambridge, Woodhead Publishing Ltd. 2005, 279-308.

11. Doyle M.P. *Campylobacter jejuni*, in *Foodborne Diseases*. Ed. Cliver D.O. London, Academic Press. 1990, 218-22.

12. Doyle M.P., Roman D.J. Growth and survival of *Campylobacter fetus* subsp. *jejuni* as a function of temperature and pH. *Journal of Food Protection*, 1981, 44 (8), 596-601.

13. Austin J. *Clostridium botulinum*, in *Food Microbiology: Fundamentals and Frontiers*. Eds. Doyle M.P., Beuchat L.R., Montville T.J. Washington DC, ASM Press. 2001, 329- 49.

14. Novak J., Peck M., Juneja V., Johnson E. *Clostridium botulinum* and *Clostridium perfringens*, in *Foodborne Pathogen. Microbiology and Molecular Biology*. Eds. Fratamico P.M., Bhunia A.K., Smith J.L.. Great Britain, Caister Academic Press. 2005, 383-408.

15. Kim J., Foegeding P.M. Principles of Control, in *Clostridium botulinum: Ecology and Control in Foods*. Eds. Hauschild A.H.W., Dodds K.L. New York, Marcel Dekker. 1993, 121-76.

16. Lund B.M, Peck M.W. *Clostridium botulinum*, in *The Microbiological Safety and Quality of Food, Volume 2*. Eds. Lund B.M., Paird-Parker T.C., Gould G.W. Gaithershurg, Aspen Publications. 2000, 1057 – 1109.

17. Dodds, K.L. *Clostridium botulinum*, in *Foodborne Disease Handbook, Volume 1: Diseases Caused by Bacteria*. Eds. Hui Y.H., Gorman J.R., Murrell K.D., Cliver D.O. New York, Marcel Dekker. 1994, 97-131.

18. Hauschild A.H.W. *Clostridium botulinum*, in *Foodborne Bacterial Pathogens*. Ed. Doyle M.P. New York, Marcel Dekker. 1989, 111-89.

19. Lund B.M., Notermans S.H.W. Potential hazards associated with REPFEDS, in *Clostridium botulinum: Ecology and Control in Foods*. Eds. Hauschild A.H.W., Dodds K.L. New York, Marcel Dekker. 1993, 279-303.

20. Betts G.D., Gaze J.E. Growth and heat resistance of psychrotrophic *Clostridium botulinum* in relation to 'sous vide'. *Food Control*, 1995, 6 (1), 57-63.

21. Wrigley D.M. *Clostridium perfringens*, in *Foodborne Disease Handbook, Volume 1: Diseases Caused by Bacteria*. Eds. Hui Y.H., Gorham J.R., Murrell K.D., Cliver D.O. New York, Marcel Dekker. 1994, 133-67.

22. Labbe R., Juneja V.K. *Clostridium perfringens* gastroenteritis, in *Foodborne Infection and Intoxication*. Eds. Riemann H.P, Cliver D.O. London, Elsevier. 2006, 137-64.

23. Labbe R. *Clostridium perfringens*, in *Foodborne Bacterial Pathogens.* Ed. Doyle M.P. New York, Marcel Dekker. 1989, 191-243.

24. Lund B.M. Foodborne disease due to *Bacillus* and *Clostridium* species. *Lancet,* 1990, 336 (8721), 982-6.

25. Johnson E.A. *Clostridium perfringens* food poisoning, in *Foodborne Diseases.* Ed. Cliver D.O. London, Academic Press. 1990, 229-40.

26. McClane B.A. *Clostridium perfringens*, in *Food Microbiology: Fundamentals and Frontiers.* Eds. Doyle M.P., Beuchat L.R., Montville T.J. Washington D.C., ASM Press. 2001, 351-82.

27. Iversen C., Lehner A., Mullane N., Marugg J., Fanning S., Stephan R., Joosten H. Bidlas E., Cleenwerck I. The taxonomy of *Enterobacter sakazakii*: proposal of a new genus *Cronobacter* gen. nov. and descriptions of *Cronobacter sakazakii* comb. nov. *Cronobacter sakazakii* subsp. *sakazakii*, comb. nov., *Cronobacter sakazakii* subsp. *malonaticus* subsp. nov., *Cronobacter turicensis* sp. nov., *Cronobacter muytjensii* sp. nov., *Cronobacter dublinensis* sp. nov. and *Cronobacter genomospecies 1. BMC Evolutionery Biolog,.* 2007; 7 (64)

28. Iversen C., Lehner A., Mullane N., Marugg J., Fanning S., Stephan R., Joosten H. Identification of "*Cronobacter*" spp. (*Enterobacter sakazakii*). *Journal of Clinical Microbiology,* 2007, 45 (11), 3814–6.

29. Grant I.R., Houf K., Cordier J.-L., Stephan R., Becker B., Baumgartner A. *Enterobacter sakazakii. Mitteilungen aus Lebensmitteluntersuchung und Hygiene,* 2006, 97 (1), 22-7.

30. Baxter. P. Have you heard of *Enterobacter sakazakii? Journal of the Association of Food and Drugs Officals,* 2005, 69 (1), 16-7.

31. Breeuwer P., Lardeau A., Peterz M., Joosten H.M. Desiccation and tolerance of *Enterobacter sakazakii. Journal of Applied Microbiology,* 2003, 95 (3), 967-73.

32. Deseo J. Emerging pathogen: *Enterobacter sakazakii. Inside Laboratory Management,* 2003, 7 (3), 32-4.

33. Kim H., Ryu J.-H., Beuchat L.R. Survival of *Enterobacter sakazakii* on fresh produce as affected by temperature, and effectiveness of sanitisers for its elimination. *International Journal of Food Microbiology,* 2006, 111 (2), 134-43.

34. Advisory Committee on the Microbiological Safety of Food. *Report on verocytotoxin-producing Escherichia coli. London,* HMSO. 1995.

35. Meng J., Doyle M.P., Zhao T., Zhao S. Detection and control of *Escherichia coli* O157:H7 in foods. *Trends in Food Science and Technology,* 1994, 5 (6), 179-85.

36. Buchanan R.L., Bagi L.K. Expansion of response surface models for the growth of *Escherichia coli* O157:H7 to include sodium nitrite as a variable. *International Journal of Food Microbiology,* 1994, 23 (3, 4), 317-32.

37. International Commission on Microbiological Specifications for Foods. Intestinally pathogenic *Escherichia coli*, in *Microorganisms in Foods, Volume 5: Microbiological Specifications of Food Pathogens.* Ed. International Commission on Microbiological Specifications for Foods. London, Blackie. 1996, 126-40.

38. Meng J., Doyle M.P. Microbiology of Shiga-toxin-producing *Escherichia coli* in foods, in *Escherichia coli O157:H7 and Other Shiga Toxin-producing E. coli Strains*. Eds. Kaper J.P., O'Brien A.D. Washington D.C., American Society for Microbiology. 1998, 92-108.

39. Glass K.A., Loeffelholz J.M., Ford J.P., Doyle M.P. Fate of *Escherichia coli* O157:H7 as affected by pH or sodium chloride and in fermented, dry sausage. *Applied and Environmental Microbiology*, 1992, 58 (9), 2513-6.

40. Walker S.J., Archer P., Banks J.G. Growth of *Listeria monocytogenes* at refrigeration temperatures. *Journal of Applied Bacteriology*, 1990, 68 (2), 157-62.

41. *Listeria monocytogenes*, in *Food Microbiology - An Introduction*. Eds. Montville T.J., Matthews K.R. Washington, ASM Press. 2005, 173-88.

42. International Commission on Microbiological Specifications for Foods. *Listeria monocytogenes*, in *Microorganisms in Foods, Volume 5: Microbiological Specifications of Food Pathogens*. Ed. International Commission on Microbiological Specifications for Foods. London, Blackie. 1996, 141-82.

43. Bradshaw J.G., Peeler J.T., Corwin J.J., Hunt J.M., Twedt R.M. Thermal Resistance of *Listeria monocytogenes* in dairy products. *Journal of Food Protection*, 1987, 50 (7), 543-4.

44. Ryser E.T., Marth E.H. *Listeria, Listeriosis and Food Safety*. New York, Marcel Dekker. 2007.

45. Kim C.J., Emery D.A., Rinke H., Nagaraja K.V., Halvorson D.A. Effect of time and temperature on growth of *Salmonella enteritidis* in experimentally inoculated eggs. *Avian Disease*, 1989, 33, 735-42.

46. Wallace G.I. The Survival of Pathogenic Microorganisms in Ice Cream. *Journal of Dairy Science*, 1938, 21 (1), 35-6.

47. International Commission on Microbiological Specifications for Foods. Salmonellae, in *Microorganisms in Foods, Volume 5: Microbiological Specifications of Food Pathogens*. Ed. International Commission on Microbiological Specifications for Foods. London, Blackie. 1996, 217-64.

48. D'Aoust, J.-Y. *Salmonella*, in *Foodborne Bacterial Pathogens*. Ed. Doyle M.P. New York, Marcel Dekker. 1989, 327-445.

49. Gustafson J., Wilkinson B. *Staphylococcus aureus* as a food pathogen: staphylococcal enterotoxins and stress response systems, in *Understanding Pathogen Behaviour Virulence, Stress Response and Resistance*. Ed. Griffiths M. Cambridge, Woodhead Publishing, 2005, 331-57.

50. Reed G.H. Foodborne illness (Part 1): Staphylococcal ("Staph") food poisoning. *Dairy, Food and Environmental Sanitation*, 1993, 13 (11), 642.

51. Bergdoll M.S, Lee Wong A.C. Staphylococcal intoxications, in *Foodborne Infections and Intoxications*. Eds. Riemann H.P., Cliver D.O. London, Academic Press. 2005, 523- 62.

52. Jay J.M., Loessner M.J., Golden D.A. Staphylococcal gastroenteritis, in *Modern Food Microbiology*. Eds. Jay J.M., Loessner M.J., Golden D.A. New York, Springer Science. 2005, 545-66.

53. Bergdoll M.S. Staphylococcal Food Poisoning, in *Foodborne Disease*. Ed. Cliver D.O. London, Academic Press. 1990, 85-106.

54. Stewart G.C. *Staphylococcus aureus*, in *Foodborne Pathogens: Microbiology and Molecular Biology*. Eds. Fratamico P.M., Bhunia A.K., Smith J.L. Wymondham, Caister Academic Press. 2005, 273-84.

55. Nesbakeen T. *Yersinia enterocolitica*, in *Foodborne Infections and Intoxications*. Eds. Reimann H.P., Cliver D.O. Oxford, Elsevier. 2006, 289-312.

56. Nesbakeen T. *Yersinia enterocolitica*, in *Emerging Foodborne Pathogens*. Eds. Motarjemi Y., Adams M. Cambridge, Woodhead Publishing. 2006, 373-405.

57. International Commission on Microbiological Specifications for Foods. *Yersinia enterocolitica*, in *Microorganisms in Foods, Vlume 5. Microbiological Specifications of Food Pathogens*. Ed. International Commission on Microbiological Specifications for Foods. London, Blackie. 1996, 458-78.

58. Feng P., Weagant S.D. *Yersinia*, in *Foodborne Disease Handbook, Voume 1. Diseases Caused by Bacteria*. Eds. Hui Y.H., Gorham J.R., Murrell K.D., Cliver D.O. New York, Marcel Dekker. 1994, 427-60.

59. Robins-Browne, R.M. *Yersinia enterocolitica*, in *Food Microbiology: Fundamentals and Frontiers*. Eds. Doyle M.P., Beuchat L.R., Montville T.J. Washington D.C., ASM Press 1997, 192-215.

CONTACTS

Addresses of Trade Associations and Professional Bodies

The Dairy Council
Henrietta House
17-18 Henrietta Street
Covent Garden
London
WC2E 8QH
United Kingdom
Tel: + 44 (0) 207 3954030
Fax: + 44 (0) 207 2409679
Email: info@dairycouncil.org.uk
Web site: www.milk.co.uk

Dairy Industry Federation
19 Cornwall Terrace
London
NW1 4QP
United Kingdom
Tel: + 44 (0) 207 4867244
Fax: + 44 (0) 207 4874734
Email: mail1@dif.org.uk

European Dairy Association (EDA)
14 Rue Montoyer
1000 Brussels
Belgium
Tel: + 32 25495040
Fax: + 32 25495049
Email: eda@euromilk.org
Web site: www.euromilk.org

International Dairy Federation (IDF)
Diamant Building,
Boulevard Auguste Reyers 80
1030 Brussels
Belgium
Tel: + 32 27339888
Fax: + 32 27330413
Email: info@fil-idf.org
Web site: www.fil-idf.org

Irish Dairy Industries Association (Food and Drink Industry Ireland)
Confederation House 84-86
Lower Baggot Street
Dublin 2
Ireland
Tel: + 353 1 6051560
Fax: + 353 1 6381560
Email: claire.mcgee@ibec.ie
Web site: www.fdii.ie

Scottish Dairy Trade Federation
Phoenix House
South Avenue
Clydebank
Glasgow
G81 2LG
United Kingdom
Tel: + 44 (0)141 9511170
Fax: + 44 (0) 141 9511129

American Dairy Science Association (ADSA)
1111 N. Dunlap Avenue
Savoy
IL 61874
United States of America
Tel: + 1 21 73565146
Fax: + 1 21 73984119
Email: adsa@assochq.org
Web site: www.adsa.org

Society of Dairy Technology
PO Box 12
Appleby in Westmorland
Cumbria
CA16 6YJ
United Kingdom
Tel: + 44 (0) 1768 354034
Email: execdirector@sdt.org
Web site: www.sdt.org

Other Sources of Information

Food Standards Agency
Aviation House
125 Kingsway
London
WC2B 6NH
United Kingdom
Tel: + 44 (0) 207 2768000
Fax: + 44 (0) 207 238 6330
Emergencies only: + 44 (0) 207 270 8960
Email: helpline@foodstandards.gsi.gov.uk
Web site: www.foodstandards.gov.uk

Department for Environment Food and Rural Affairs (Defra)
Nobel House
17 Smith Square
London
SW1P 3JR
United Kingdom
Tel: + 44 (0) 207 2386000

Fax: + 44 (0) 207 2382188
Email: helpline@defra.gsi.gov.uk
Web site: www.defra.gov.uk

Institute of Food Research (IFR)
Norwich Research Park
Colney lane
Norwich
NR4 7UA
United Kingdom
Tel: + 44 (0) 160 3255000
Fax: + 44 (0)160 3507723
Web site: www.ifr.ac.uk

Institute of Food Science and Technology (IFST)
5 Cambridge Court
210 Shepherds Bush Road
London
W6 7NL
United Kingdom
Tel: + 44 (0) 207 6036316
Fax: + 44 (0) 207 6029936
Email: info@ifst.org
Web site: www.ifst.org

Health Protection Agency (HPA) Centre for Infections
61 Colindale Avenue
London
NW9 5EQ
United Kingdom
Tel: + 44 (0) 208 2004400
Fax: + 44 (0) 208 2007868
Web site: www.hpa.org.uk

Chilled Food Association
PO Box 6434
Kettering
NN15 5XT
United Kingdom
Tel: + 44 (0) 1536 514365
Fax: + 44 (0) 1536 515395
Email: cfa@chilledfood.org
Web site: www.chilledfood.org

Food and Drink Federation (FDF)
6 Catherine Street
London
WC2B SJJ
United Kingdom
Tel: + 44 (0) 207 8362460
Fax: + 44 (0) 207 8360580
Email: generalenquiries@fdf.org.uk
Web site: www.fdf.org.uk

Useful Web Sites

Gateway to Government Food Safety Information (US)
  http://www.FoodSafety.gov/

World Health Organization (WHO): Food safety programmes and projects
  http://www.who.int/foodsafety/en/

European Commission: Activities of the European Union - Food Safety
  http://europa.eu/pol/food/index_en.htm

Centre for Disease Control and Prevention (CDC) (US)
  http://www.cdc.gov/foodsafety/

Food Safety Authority of Ireland
  http://www.fsai.ie/

Food Science Australia
  http://www.foodscience.csiro.au/

International Association for Food Protection
  http://www.foodprotection.org/

Institute of Food Technologists
  http://www.ift.org/

Grocery Manufacturers Association
  http://www.gmaonline.org/

Royal Society for Public Health (UK)
  http://www.rsph.org.uk/

Society of Food Hygiene and Technology (UK)
  http://www.sofht.co.uk/

INDEX